It's Not Easy Being Green

It's not easy

Being Green

One family's journey towards eco-friendly living

Dick Strawbridge

BBC
BOOKS

To Brigit, James and Charlotte –
they put up with an awful lot

Photography by Nick Smith
(www.nicksmithphotography.com)
with the exception of pages 69, 148,
212 © Dick Strawbridge

This book is published to accompany the television
series *It's Not Easy Being Green* made for BBC2 by
Documentaries and Contemporary Factual, BBC Bristol.
Series Producer: Lynn Barlow
Executive Producer: Julian Mercer

First published by BBC Books in 2006
Reprinted 2006
BBC Worldwide Limited
80 Wood Lane
London W12 0TT

ISBN 10: 0 563 49346 1
ISBN 13: 978 0 563 49346 4

Commissioning Editor: Vivien Bowler
Project Editor: Eleanor Maxfield
Art Director and Designer: Lisa Pettibone
Illustration: Lisa Pettibone
Production Controller: Kenneth McKay

Set in Egyptian and Frutiger

Colour origination and printing by
Butler and Tanner Ltd, Frome, England

PUBLISHER'S NOTE *The book contains imperial
measurements throughout, except in those instances
where the metric unit is in more common usage.*

Contents

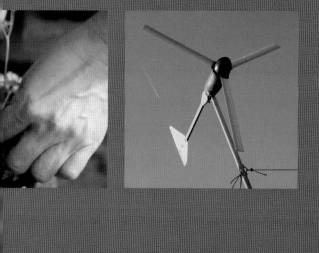

Introduction

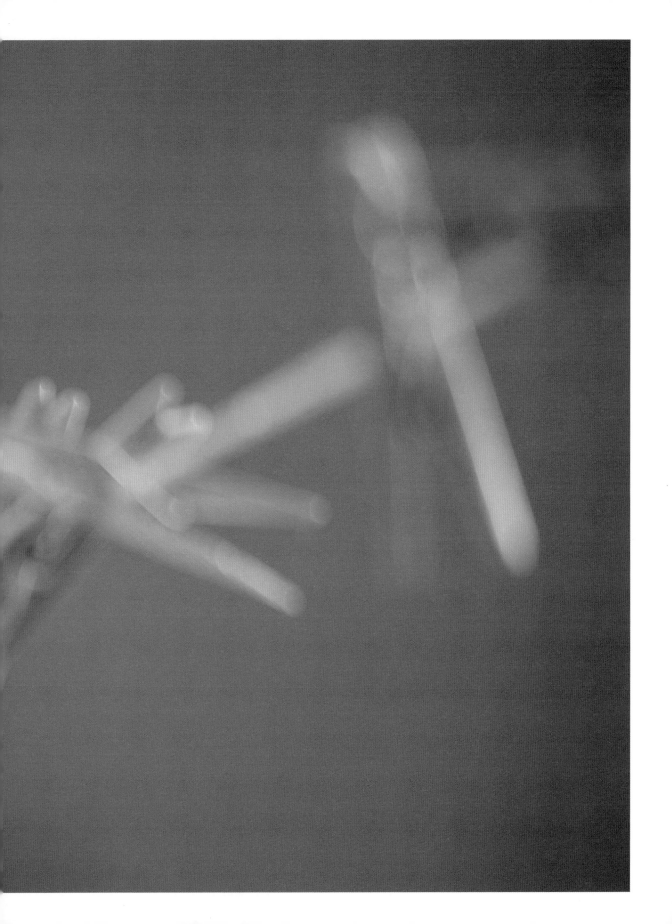

The Strawbridge family objective (at the start!): to live a 21st-century lifestyle, but to produce little or no waste and to remove our dependence upon fossil fuels

What it means to be green

The Strawbridge family were very comfortable and happy in our spacious house in the village of West Malvern, Worcestershire. Yet we packed our bags and moved to a derelict farmhouse in Cornwall without electricity, heating, water or loos. Cornwall is a beautiful place, and living by the sea sounds idyllic to many, so our choice of location isn't so surprising. Losing all our comforts, and setting ourselves the task of completely rebuilding a 300-year-old property, might take a little more explaining.

Actually, it's not all that complicated. It dawned on us that we wanted to live in an environmentally friendly way. 'Dawned on us' gives the impression that we made a sudden, collective, coherent decision, but it wasn't really like that. I suppose the family began changing when the children came home from primary school and started lecturing my wife Brigit and I on turning off the tap while we brushed our teeth and putting only enough water in the kettle for the number of cups of tea we were going to make. Their logic was unassailable and, from there, a series of small actions and a gradual awakening led us to the point where, some years down the line, we decided to take action rather than just talk about green issues.

The challenge we set ourselves was to live a sustainable lifestyle without giving up the advantages of modern, 21st-century Britain – we are definitely not 'eco-warriors', and we were not prepared to forgo our creature comforts for long. Our impending house move would provide us with the opportunity to be greener, but we still had to determine exactly how far we were prepared to go. The first step was to be a reduction of our 'eco-footprint'. This grand statement doesn't quantify much, but it does allow us to stand on the moral high ground with a rather impressive flag

saying, 'Look at us, we are green – feel guilty the rest of you, destroyers of the planet!'

As I explained, we decided to move house, but there was a lot more to it than that. There is a general growing awareness of green issues and government initiatives to reduce waste that have led to increased public participation in sustainability.

Like so many families, we didn't understand exactly what was required of us to be truly green so we decided it was time to start researching and doing something. After a fair amount of deliberation we agreed we would attempt to jump in at the deep end, and try to start really living the green life! Because our efforts were televised by the BBC, we were fortunate enough to be able to try to do everything at once rather than slowly changing. That doen't mean that small steps aren't worth taking, but, as we discovered, the title of the series and this book was very apt: 'It's Not Easy Being Green'.

There are several mantras you discover when you embark on a journey into green living. The most important is:
Reduce + Reuse + Recycle

Reduce

I'm a very simple, if somewhat cynical, individual, but it is so obvious that the best and foremost way of being green is to reduce how much you use. I have a major problem getting my head around how much 'tat' people (myself and my family included) are suckered into buying. We are a consumer society, which is not necessarily awful, but we appear to have lost the ability to resist the marketing forces that pull us along buying all before us. That said, I like new things, I like clever things and I like shopping (though I must qualify the latter by explaining that this, for me, tends to be a focused activity rather than a ramble around). There is the added advantage that reducing the amount you buy allows you to justify spending that little bit more on what you really need and want, and if you take this to its logical conclusion you should have more left in the bank as a result of being greener.

Reuse

Giving existing possessions and goods a new lease of life is the idea, but it does need to involve minimum expenditure of energy on reprocessing or modifying – there are lots of examples of innovative reuse later in the book. I particularly like the old radiator that we turned into a solar shower. At this stage I have to give a health warning: my wife Brigit has decided that it is essential to visit car boot sales as her main reuse initiative, and consequently we gain a carload of goodies every Sunday morning that are being given a new home and a new lease of life. I'm not sure why she selectively fails to remember 'reduce' …

Recycle

I suppose recycling is the facet of being green that most people are aware of and attempt to do, though it can be tiresome organizing bins and boxes. Right up front I have to say that recycling is essential. We need to encourage the recycling industry by stimulating the market for recycled goods and providing the raw materials. When I first went to Germany twenty years ago, it was illegal to put glass bottles and paper in the dustbins – they had to be separated or you could be fined. It has taken successive governments here a long time to do

12

anything, but initiatives have now led to increased public participation in activities such as recycling. Anyone visiting the local council 'dump' can see recycling going on. My non-scientific study of the cars driven to the dump shows that most are less than five years old and their occupants are solid middle-class citizens doing their duty. Unfortunately, they are probably using more energy driving their recyling to the dump than will be saved through the process.

For a glass jar to be turned into a wine glass it has been taken from home to the bottle bank, then taken to a processing plant, processed to a usable form, taken to a wine-glass manufacturer, manufactured, taken to wholesalers, then to a retailer, bought and finally taken home (energy! fossil fuels! pollution!). This is better than just throwing it away, but why not reuse the jar – stick some spices/jam in it? Better still, don't buy it in the first place – get your honey from a local supplier who will fill an existing jar. But recycling is better than throwing away.

So after a lot of deliberation we decided that we should set our bar rather high, and, being thoroughly British, we know that it is in the endeavour that one grows rather than in the achieving of the goal.

HOW BIG IS YOUR ECO-FOOTPRINT?

Environmentalists often explain sustainability by referring to our ecological footprint: the amount of land, water and other natural resources required to support our lifestyle. In 2001, humanity's ecological footprint exceeded global bio-capacity by 21 per cent. In other words, we are now using natural resources faster than they can be replaced, something that has been occurring since the 1980s. And if everyone in the world had the lifestyle that we take for granted in the UK, it would take eight Earths to support us all.

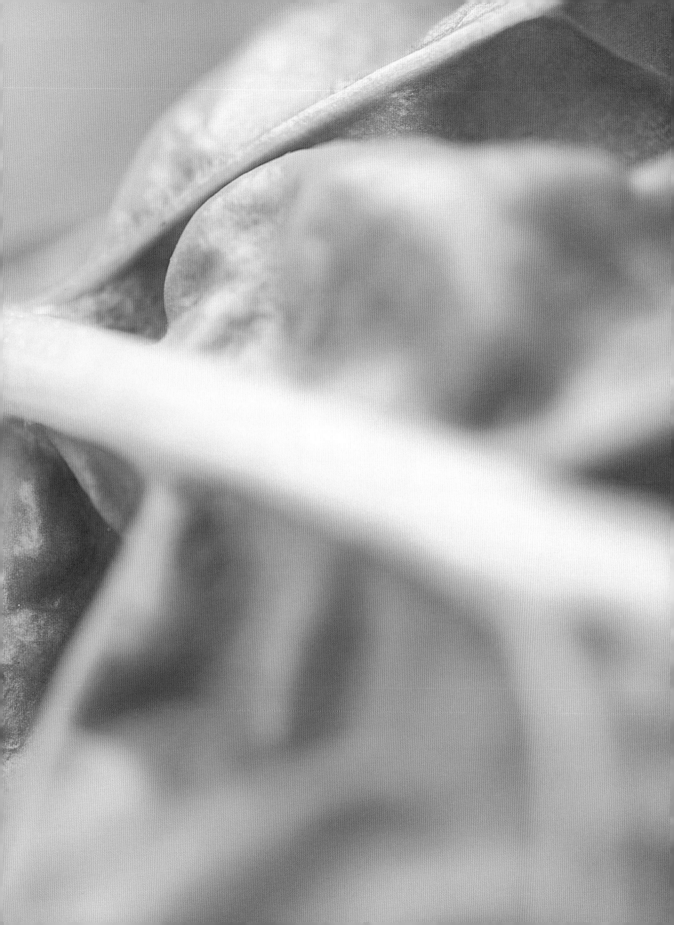

Of the family, I'm probably the simplest to understand. My wife Brigit (bottom right), James (far right) and Charlotte (top left) all pick on me because I am rather conventional in my outlook.

Why the television cameras?

We reluctantly concluded that we needed to move as our house in West Malvern did not have the facilities to produce the opportunites we neeed to 'go greener' – it did not have a workshop, or the space to build one, for example. As a family there were two major considerations: when and where? With James and Charlotte finishing school, doing the gap year 'thing' and heading off for university, the time seemed right. Theoretically, Brigit and I now had less responsibility and could do what suited us. We toyed with moving to continental Europe, mainly because we could afford to buy land and property without having to service a painful mortgage, but we like Britain so decided to narrow our search.

During my military career I was surrounded by barrack walls with notices such as: 'Careless talk costs lives' and 'Walls have ears'. I suppose I should have been more careful, but discussions of our plans made it back to the ears of the development team at BBC Bristol, via our good friend (and director) Steve Rankin – cheers, mate! As it happened, they were keen to make a green series, and a quick chat and a taster video later, *It's Not Easy Being Green* was launched. To set the record straight, I should point out that we all had our misgivings about exposing the soft underbelly of family life with the Strawbridges. Rather than a 'quick chat' it was in fact a serious family discussion drawn out over many weeks, but it led to the conclusion that, with the assurances we had from the team at BBC Bristol, the project would be worthwhile and could make a real difference if viewers learned from watching – so we decided we were prepared to allow the cameras into our home.

We set ourselves a completely unrealistic timescale and then had to stick to it. The people we met who were doing similar things had been slowly evolving and growing into their lifestyle. We started filming in spring 2005 and the crews left, after a cracking New Year's party, on 1 January 2006. We came a very long way very quickly (which I liked), but life at New House Farm was frantic. You get less done on a filming day than you would without cameras. On the other hand, the BBC is enormously well respected, meaning that in lots of areas we got more support than the average customer. It's a bit of a pity the BBC doesn't do advertising. Some judicious product placement would have saved us a fortune!

I suppose this is as good a time as any to point out that, contrary to common belief, the BBC didn't pay for my house to be done up or for the eco-projects we undertook. I was paid as a presenter – everything else we did was down to us working within our budget. This meant that we were prepared to adopt only green ideas and 'gadgets' that truly delivered a benefit.

The team

Brigit, my wife, is 46, and since she recently ran the London Marathon for the charity Children with Leukaemia I suppose she could now be classed as an athlete. She quotes long passages of *The Hobbit* at the most obscure times, is a great healer/listener and is definitely away with the fairies. Brigit has a spirituality which, I must admit, I struggle to understand and she is receptive to everything 'alternative'. Throughout our adult life we have moved frequently, but the last 11 years have been spent in Malvern, where Brigit has built a wide network of friends, so the move to Cornwall was a massive wrench for her. That said, she cares very deeply about nature and the planet and is probably the main driving force behind the project.

James is 21 and a long-haired history student at York University. He is artistic and fit, and writes poetry and books of his philosophical thoughts. From a very early age James has loved being outdoors. He has always cared about the planet, but a gap year in Nepal, conducting environmental projects with the local people, made him even more convinced that something has to be done. I think it's fairly accurate to say James has a bit of an eccentric streak …

Charlotte is 19, absolutely gorgeous (not that I'm biased), a singer-songwriter, and has just started a music degree in Newcastle. She spent six months at Havana University learning Spanish language and guitar. Like the rest of the family, Charlotte is outspoken but she is extremely passionate about justice and ethical consumerism. Maybe that's because she loves shopping?

Of the family, I'm probably the simplest to understand. Brigit, James and Charlotte all pick on me because I am rather conventional in my outlook. After I retired as a lieutenant colonel, I moved to industry where I was a successful programme manager known for getting jobs done to time and on budget. I was also a trouble-shooter, and it is fair to say my sensitive side is a little under-developed. Apparently, I have some issues with my *Yin* and *Yang* not being in balance! For me, the most exciting things about the project were the self-sufficiency and the engineering challenges. I also care about the cost of being green, have issues with being preached at by worthy people and think that marketing using a 'green wash' (making theoretically true but vastly inflated claims on the green credentials of a product – and thereby justifying an increased price) means that there is very little that can be taken at face value.

We also have two friends working with us. Jim Milner is a 28-year-old research scientist from QinetiQ (www.qinetiq.com) who is experienced in eco-engineering. He's our answer to Charlie Dimmock and is even working on his pecs. If we don't know the answer to a question or if we try to source some special little widget, Jim will be found on the internet sorting it out – very useful! Anda Phillips is our permaculture expert (more later for the uninitiated) and she has been working on organic farms and

smallholdings most of her adult life – I've been too gallant to ask her age! She's inde-fatigable, working all day long, every day.

There was an awful lot to do on the house and outside. In addition to contract-ing in builders, roofers, electricians and an impressive array of experts, we press-ganged friends to come and stay. They got free bed and board by the seaside – we got their bodies during the day.

One of the advantages of having the BBC following our progress is that we also had the production team around a lot of the time. We laid down a couple of criteria before BBC Bristol selected its team. First, they had to be prepared to get stuck in with us when they were not filming (free labour!) and, probably most important, they had to have a sense of humour. Our family has been further extended by the addition of Claire Martin, Sanna Handslip, Jonny Young, Rich Whitley and a number of soundmen, most notably Nick Allison. They have shared with us the floods and sunburn and all cared as much as we do about the project – they are great.

Rather than making a gooey thank you in the acknowledgements, I'd just like to say that everyone you see in the pictures and who gets a mention has probably been woken by my early-morning singing (I find it works wonders getting people up and out to work) and has donated blood and sweat to help us.

Living the green life

The obvious thing to do would have been to buy a brand-new, state-of-the-art, environmentally sound house designed specifically to be green with all possible energy-saving devices. But, like many home-buyers, we didn't find that appealing: none of us wanted to live in a new house, no matter how clever. As a family we decided that we did not want a completely derelict house that would involve our spending months in a caravan, yurt, tipi or B&B (depending on who in the family was choosing the temporary accommodation), and it didn't seem to be a frightfully green approach to buy a perfectly serviceable property and then to 'modify' it extensively.

We needed sufficient latitude to explore how green we could be, so we drew up the parameters for our perfect property (see below).

After months of searching Herefordshire, Shropshire, most of the Midlands, Gloucestershire, Wiltshire and Somerset, we headed for Dorset, Devon and Cornwall. We had been tempted by a detached Georgian farmhouse in Shropshire with 5 acres, which we reckoned we could just about afford, but Brigit wanted to knock down half the house and she felt the village lacked soul. We probably would have bought a Victorian mill with 10 acres in Herefordshire if our initial house sale had gone through, but in hindsight, apart from oodles of water power and sufficient land, it didn't meet the rest of our requirements and would not have been right.

	WHAT WE WANTED	WHAT WE WOULD HAVE SETTLED FOR
House condition	Needs significant modernization	Good value; anything that we could move into
Size	Four bedrooms	Three bedrooms
Price	Buy, fix up and do eco-projects for less than £450,000	Maximum total budget: £500,000
Outbuildings	Several usable	Some in any condition
Location	Edge of a village with some life and community feel	Walking distance to civilization
The land	5 acres	2 acres
The stream	Good flow and a good drop across the property	Some flow

We needed sufficient latitude to explore exactly how green we could be ...

We spent several days (and nights) by the seaside and fell in love with it. As luck would have it, we found an awesome old property, in need of some serious TLC, set in a gorgeous Cornish village, half a mile from a sandy beach. As is the way with most people buying property, we knew it was too expensive and needed far too much done to it, but, being the idiots we are, we wanted it anyway. Our hearts ruled our heads, and we became the proud owners of a little bit of Cornwall called New House Farm. Did I mention it's a listed building?

New House Farm in Twyardreath didn't quite meet our stringent criteria ...

WHAT WE GOT

House	Grade 2 listed – the property is thought to go back some 300–400 years (some investigations date part of it to the 11th century). Believed to be a medieval priory located in the grounds; need an archaeologist present if digging below topsoil
Condition	Holes in the roof, lots of it falling down, no water, loos or electricity in the house
Size	Six/seven bedrooms and an amazing kitchen of 39 x 14 feet
Price	A lot more than we budgeted
Outbuildings	Four derelict/unusable barns (they do have planning permission to be converted into holiday cottages but we have no intention of going that way), an old potting shed with a leaky roof and a pole barn that is mainly waterproof
Location	Edge of a great village with lots of life and community feel. We have an award-winning butcher who makes really tasty sausages, a shop, a post office, a garage, a hairdresser and, of course, a pub with great beer. Plus the bonus of being near the seaside
The land	Set in a small south-westerly valley; 2 1/2 acres including lovely old walled garden
The stream	Our own spring, and a spring-fed stream that is not very powerful but has an impressive drop

The New House Farm site map ...

LOWER PADDOCK

SPRING

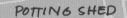

POTTING SHED

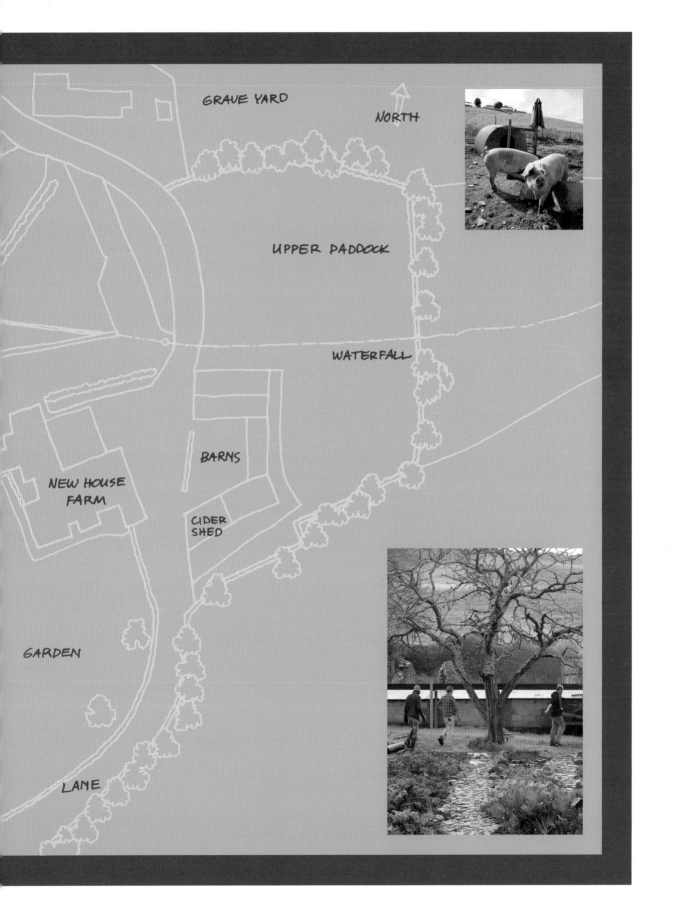

GRAVE YARD

NORTH

UPPER PADDOCK

WATERFALL

BARNS

NEW HOUSE
FARM

CIDER
SHED

GARDEN

LANE

When you make the decision to try and be greener, there is a sort of guilt that comes over you every time you pop something in the bin that you know can be recycled

Preparation

As we were about to commit all our worldly goods, and our life juices, to this project, we had to get our act together. In the army we used to swear by the seven Ps (prior planning and preparation prevents piss-poor performance), so while we were waiting for our house to sell and our purchase to go through, we started to collect information. Our metamorphosis began.

Apart from John Seymour's *Complete Book of Self-Sufficiency*, which I have had for years, and a collection of books on gardening and how to harvest from the wild, we were a bit lacking in reference material. Brigit soon sorted that out, and we are now the owners of a pretty significant library on sustainable living, all aspects of permaculture, self-sufficiency and 'eco-engineering'. You can find a list of our favourites at the back of this book. The books are great and we regularly refer to them.That said, actually doing is the important bit, and it is only now, after having owned John Seymour's book for nearly 25 years, that I have got round to doing what I've always wanted to. I could blame my military career, even my time in industry, but the bottom line is that they are excuses, not reasons. Buy the books by all means, but plan to do!

The start of the process

During the last five years in Malvern, our lifestyle had already started to change. We had begun to think more about the food we were buying, for example. Without the luxury of a vegetable patch, we had to buy most of our fresh produce. One of our first actions was to buy food without packaging and, wherever possible, produce in season. We subscribed to an organic box scheme, which was very convenient if you could get used to passing responsibility for choosing your food on to someone else. I never did quite get used to that. I also had some concerns about the 'food miles' involved. I began to realize that 'local' was more important to me than 'organic'. Locally produced organic food is the ultimate, but providing I know the source of the meat is not intensively reared and that the vegetables are not sprayed to death, I'm happy to stray away from the Soil Association logo. Brigit would still choose it over anything else.

We also had a clearout before the move. This qualifies as 'reusing' so long as you find a good home for the bits you don't want – Brigit distributed our unwanted stuff among the local charity shops. This was a definite win-win. It is good Feng Shui to declutter (apparently!) and we probably saved a removal lorry, too.

The green audit

As the production team members in BBC Bristol have a sense of humour, they thought it would be great if we had a 'green audit' on our Malvern lifestyle so that at the end of our project we could see how far we'd come. Donnachadh McCarthy, author of *Saving the Planet without Costing the Earth*, was roped in to giving us the once over. I think it's fair to say he found the day pretty stressful and very hard work! Having travelled up from London to the wilds of Worcestershire, Donnachadh took us through most of the questions in his book and scored us on how sustainable our way of living was. Things started off quite well and it was encouraging to see that we were actually fairly advanced in several areas. It started to get slightly more intense when I began to argue to get the odd extra point here or there. It didn't take long before we were in open conflict over nearly every question.

By the time Donnachadh had to leave for his train six hours later he was absolutely jiggered, but I felt that we had scored well – okay, so that wasn't the objective, but I had a great day's sport.

STOP THE JUNK

We were plagued with junk mail until Brigit got the contacts for the Mailing Preference Service. It took a bit of time to kill it, but just making the call is empowering. The MPS registration line is 0845 703 4599. The website address is www.mpsonline.org.uk. Don't just read this – make the call and stop the junk!

If you call the following number you can stop people trying to sell you something over the phone just as you settle down with a cup of tea: Telephone Preference Service registration line – 0845 070 0707 (www.tpsonline.org.uk).

Starting to sow

The delays in selling our house meant that we were going to miss the early growing season. Rather than waiting, we decided we'd start growing and worry about how to transport the plants later. Anda was well and truly up for the challenge. Very quickly we had a polytunnel full of delicate little seedlings. A polytunnel is literally a polythene covering over a tunnel frame and acts as a temporary greenhouse – ours had a heater inside. Every day it was a little chilly, Anda or I would nip up to the polytunnel to check that the heating was on. In warmer weather, as the plants grew and the completion date for our house sale was finally set, we were saved from later difficulties by the fact that a rather large chestnut tree produced sufficient shade to stop the seedlings from growing too big. When it came time to move, an offhand comment to the chaps at Ledbury Removals that we had some plants to move was met with casual indifference – apparently they moved plants all the time, so we allowed them to move absolutely everything. A good tip is to get a quote for moving plants ten weeks before your move.

The saga of the soiled bales

Not content with using pots and miscellaneous containers, we also experimented … We had to use my Audi when purchasing our straw bales – the only way to fit them in was to have the roof down; there was still straw tucked in corners when it was sold six months later.

The straw bales have to be put into black plastic bags – sounds easy, doesn't it? Have a go then. Without removing the baling twine and destroying the integrity of your bale, you then scoop out a container-shaped hollow. We used axes, hands, carving knives.

Urine is apparently a good way to get straw to decompose. Yes, you've guessed it: it was our next job to pee on the bales. I did try to get a figure for exactly how much was needed but I think the process was a lot more 'touchy feely' than scientific. It goes without saying, but boys have a real advantage during this stage.

The hollow you have made is then filled with compost. We didn't even do this the easy way. Brigit and I have had worms in a nice little wormery for years, but the liquid fertilizer it produces is so potent it needs to be diluted or it scorches plants. Anda suggested mixing it with other wholesome ingredients so we took a wheelbarrow to the bottom of the Malvern Hills to collect the choicest molehills. The soil was excellent. Unfortunately the half-mile uphill return with a wheelbarrow laden with mud nearly killed me!

However, once the hard work was over, all that was left to do was plant away, nurture and generally encourage.

Straw bales

❁ Straw bales are a relatively uncommon way of growing plants, but they are particularly good for crops you need to transport.

❁ The idea is that the plants grown in straw bales require less water. The garden bed of straw eventually turns into compost, offering additional nutrients to your vegetables and herbs.

❁ Straw bales work best for tomatoes, peppers and cucumbers, but are not so good for root vegetables such as turnips and carrots.

We're Off!

> **The first and most obvious problem we had was the roof, which had huge holes in it. We also had water coming through walls, along lintels, down chimneys and under doors**

The move

The planning phase ran until 15 June 2005, when we completed on the purchase. The cameras had followed us in our search and during the preparations to leave our ordered life in West Malvern for chaos in Twyardreath. We had taken the conventional route of selling our home and buying a property that needed to be renovated (nothing especially innovative about that). We attempted to fund the mortgage using 'green' money from ethical banks. The combination of big holes in the roof, my military background and BBC cameras hampered us, but with the help of Andy Hopper, a mortgage consultant based in London, we did find a good company keen to lend us sufficient money to get started.

I'd always thought Brigit was smart as well as gorgeous, and I was proven right when she announced that the only permaculture course she could take was during the week before and after our move. I think I was quite chilled about Brigit's plan, mainly because it probably served me right as Brigit used to handle all our moves when I was in the army (20 moves in 23 years). Charlotte didn't take it quite so well. She had the unenviable task of packing and unpacking with me. James stayed well out of it at university. Having allowed Brigit to do some pre-packing, and having convinced her that we were up to the task, we got on with it.

My memory might be a bit selective, but Charlotte and I did a sterling job. We moved everything, we didn't break much and we sort of knew where it was when we arrived in Cornwall. (Our friends John, Dean, Phil and Steve from Ledbury Removals stepped in again as they had been looking forward to an opportunity to move our big, old, ironed-framed piano.) Truth be told, I seem to remember a couple of heated exchanges, but that must have been all Charlotte's fault ... I think the combination of Brigit's absence, leaving somewhere we all loved, the cameras and moving to a near-derelict house probably conspired to make it 'interesting'.

The Friday before our move, Jim Milner and I took a gallop down to Cornwall where the vendors, who were great throughout the purchase, gave us permission to install a loo and connect up some water and electricity. After a dash of 250 miles each way and a long day, we managed to get water into the house, put a stopcock, a new loo and sink in the utility room and get a loo upstairs. An upstairs loo seems pretty normal. However, our version was in the middle of a rather large room (14 × 25 feet) with a door at each end, neither of which locked particularly well. The girls never really took to our large loo, and it wasn't until we curtained it off and put a bath in there that they decided it was worth visiting.

After a weekend with Brigit back to help, and two days with the packers, we headed south with three removal lorries, two cats and Charlotte's Swedish boyfriend, Jens. I think it's fair to say that the cats had more room than we did. We left behind lots of friends and security, but as we headed down the M5 I started smiling inanely, and crossing the Tamar made me positively beam.

The next two and half days are now a bit of a blur, but somehow we put our worldly goods into the house, barns and anywhere dry and by Friday night, when Brigit turned

up, we looked as if we lived there. We even managed to recommission the old immersion heater (well done, Jim) and installed a bath next to the loo in the large room.

As Brigit's welcoming committee was sitting waiting for her, refusing to do any more, the general opinion was that the house looked pretty good. We were living in Cornwall by the seaside, we had some electricity, a washing machine, a cooker, lights, loos and even a bath. When Brigit arrived after a four-hour journey from Gloucestershire, the family had officially arrived in Cornwall and the fun could begin.

In places the walls are thick (up to 5 feet) and are made of a sandwich of cob, stone and brick

Layers of history

New House Farm is made of lots of materials and has been added to a bit at a time. We have yet fully to investigate the history of the property, but even the untrained eye can see how the layers have built up. A little research and conversations with local historians have confirmed that this is a real jewel of a home/plot. Somewhere under our land, or very close indeed, lie the remains of a medieval priory that was significant from the 11th to the 16th century. We have been convinced that part of the house dates back to that period.

In places the walls are thick (up to 5 feet) and are made of a sandwich of cob, stone and brick. For the uninitiated, 'cob' is an amazing mud, stone and miscellaneous debris mixture that is made into a sort of stodgy mud pie and used to build walls. At first, I found the idea of building in mud a little hard to accept. However, the thermal properties are great, it is widely available and it is 'green'. But be warned: it doesn't like getting wet, takes ages to dry out and needs to breathe, which probably makes it the wrong material to have when your roof leaks.

Priority work

Our first big task was to put on a new roof and fix all the roof timbers. However, we also had some high-priority structural

work to address: a couple of walls were not actually joined together, a very large chimney was at a rather unusual angle and our cob walls were exposed, which is not a good thing.

We also decided that we needed to record our successes and (heaven forbid) failures, so the New House Farm diary was started. Every day we record the tasks we intend to do, our achievements and any issues that arise. Everyone staying at the farmhouse is permitted to write down their thoughts and comments – it all sounds very democratic, but I own the pencil so I do maintain some control.

Having got a couple of quotes for the roof work, we had booked people to commence the week after we moved in. At our first on-site meeting with Colin (Colin Marshall Scaffold), Ross (Forrester Roofing) and Roger (Gilbert and Goode Builders), I was somewhat concerned by the fact that everyone found our problems amusing and that the general opinion was that we had our work cut out for us. And it was only after we 'saved' the house by repairing the roof that we could tackle the 'smaller' problems like plumbing, wiring and reversing the years of neglect, room by room. At every opportunity we intended to renovate using sustainable materials.

We had a cunning plan: as our highly qualified contractors conducted the essential work, the rest of our team would make

35

a concerted effort to get the majority of our planting done by the summer solstice, after which we could get on to less skilled tasks. Brigit and Anda believed it would make a huge difference to our productivity – something about biorhythms? I went along with it and convinced the blokes that the girls had some as yet unexplained logic that would probably conclude with us dancing round naked pretending to be druids. There was a distinct lack of dancing/nudity but the gardening was successful so they obviously got something right.

Noticeably absent from our list of priority work was any eco-engineering. With the house and land in such a state, Jim and I had come to terms with the fact that we couldn't play in our workshop until there was some degree of civilization in our life. It was a fair cop and we were very patient.

The roof

Colin's lads didn't mess around – we were completely encased in scaffold by the end of our first full week in Cornwall. We had been warned that if a Cornish contractor promised to do something 'directly' it would never happen. The cameras obviously focused the minds of those we worked with, but Brigit and I had made a decision not to try for reductions with promises of fame, fortune and ladies throwing their underwear at the workers, but instead simply to ask them to ensure they met their deadlines, to do what they said they would do when they said they would do it. I am happy to report that everyone we had dealings with was first class and as good as their word – they were stars!

That said, there was a couple of contractors who thought that, as there were cameras around, the BBC must be paying and that it was time to reclaim every licence fee they had ever stumped up for. As I've mentioned, the BBC wasn't paying for anything; I didn't have a sense of humour; and they didn't get any work.

Forrester Roofing was on as soon as the first scaffolding planks were up. Darren and Adam were a part of our life for the next two months and they launched straight into taking off the old roof, which was the most amazing mixture of ancient slate, bitumen, cement, asbestos and

miscellaneous goop. (For those without an engineering background, goop is a mucky, sticky, rather unpleasant substance that seems to hold things together but has a tendency to let water through.) It took only a couple of days, but as they peeled off the mess, the Gilbert and Goode builders were following behind them making good the woodwork in the roof. Steve, the rather eccentric master chippy, was like a whirl-wind. He made short work of the trusses, purlins and batons. The only time he stopped to draw breath was when he found something really extraordinary. First, there was a small window with an ancient paned frame in the middle of a wall, which had simply been covered when the walls were

thickened. Then came the best part of a tree trunk that formed the main support for one of the hips. It had been unceremoniously propped up in the ceiling of Charlotte's room hundreds of years ago and was the reason for a rather scary bowing effect. It was impressive, but I must admit what I most wanted to know was whether it would cost me more to replace it or to leave it. Steve worked round it and it is still there in all its glory, though it's not load-bearing any more. We harbour hopes that Charlotte's ceiling will miraculously creep back into shape.

Slate

One of our first major decisions was what slate to use. We couldn't afford the local Cornish Delabole slate, which was three times the price of any other. We were not allowed to use the slate that is made from recycled materials as the building is listed and the finish was not acceptable (though Ardesia slates might be of interest to anyone without restrictions www.e-b-d-uk.com). That left two viable options: Brazilian river slate and Spanish Bretona. The Brazilian slate was the cheapest, but we were not sure about the colour, and both Brigit and I had an uneasy feeling about where it came from and exactly how ethical it was to buy slate from the other side of the world. The Bretona slate came from within the EU and was an acceptable colour but it was nearly a thousand pounds more expensive. I would be hard pushed to say whether it was the colour or the source

that encouraged us to choose the Bretona, but it was duly chosen, and as it went on it transformed our moth-eaten old property.

As the Victorian part of the house still had life in its roof, we chose not to replace it. It needed to be patched and repaired in a couple of places, but that was trivial compared to the main part of the roof. From the inside it was disconcerting to see the sky through the gaps in the slates, but Darren and Adam managed to patch and realign sufficiently to make us weathertight.

The renovation of the roof uncovered a plethora of chimneys. We have six stacks of a possible thirteen; they seemed to pop up on every wall and were a legacy from an era when every room had its own fire. Brigit reckoned that while we had the roof uncovered was a good time to get the planners round to discuss some of our issues

and to update them. I must admit that my often-quoted 'it's better to beg forgiveness than to ask permission' attitude made me wary of exposing too much detail, but Brigit was right. The visit was a success. Actually, I think the planners were happy to have some-one silly enough to try to save the house while living there and they were completely behind us. A couple of times we ended up with the most aesthetically pleasing instead of the cheapest solution and I couldn't help feeling that Brigit was conspiring with them – I never stood a chance!

After weeks of activity on the roof we finally had the 'topping-out' ceremony. I'd never heard of having a celebration on completion of a roof, but someone thought it was a good idea, so as the last slate was laid there was a gathering on the scaffold. Champagne at 20 feet above ground is

From the inside it was disconcerting to see the sky through the gaps in the slates ...

probably contrary to every Health and Safety rule, but it tasted great and it was worth celebrating that we were finally waterproof.

Guttering

It was bad enough having to replace most of the roof, but we also needed to put up new guttering. The lovely, very effective new roof had created a most efficient way of channelling water so that it gushed all over anyone walking round the outside of the house. Yet again, the listed status of the house would cost us more money. Next to no cast iron survived being stripped from the fascia boards – it was rusty and rotten. ('Ogee profile' for those who need to know – it's funny what you learn as you undertake a project like an old house.) After some research we discovered that it was possible to get cast aluminium and that it was acceptable because it looked exactly the same as the cast iron. It also had the advantage of much lower maintenance overheads and

if it was ever replaced could be easily recycled. Several hours on the internet, lots of conversations with fitters and suppliers later, and we had found ARP, Aluminium Roofline Products, in Leicester (www.arp-ltd.com). They know their onions and were extremely competitive. I even convinced them to let John Hemphill, who brought our order to us, stay over and show James and his mate Henry how to put it up. John entered into the spirit of things at New House Farm and before he left the three of them had broken the back of the guttering. We now have some smart long-life gutters to colect the water off our roof. The guttering wasn't cheap, but is starting to pay for itself as we use it for rain harvesting.

And so finally, after I'd spent a couple of weeks filming the *Scrapheap Challenge* television series for Channel 4, the scaffolding could come down and we could see the shape of the house for the first time since we had moved in – the whole thing looked great and the roof was gorgeous.

Rain harvesting: the sums

- ◆ We have approximately 230 square metres of roof.
- ◆ There is approximately 1 metre of rain a year in Cornwall. (I'd like a recount – it feels like a lot more!)
- ◆ We pay the water board £3.35 per cubic metre.
- ◆ If instead of buying water we used the water on our roof, it would save us £770.50.

THE SOLAR SHOWER

Before we installed a bath, the solar shower was our only washing facility. Theoretically leaving a black plastic bag lying in the sun for a couple of hours would be sufficient to warm enough water for a couple of showers. However, we had a few problems. There were four of us – me, Charlotte, Jim and Anda – plus four removal men and between two and four BBC crew.

It wasn't frightfully sunny.

Most of us wanted a shower in the morning, which is a challenge for even the most efficient solar shower. In fact, even weeks later, when we had a much bigger solar shower, it still didn't really do the job for early-morning showers – there is a gap in the market for a lunar shower, but I think I'll leave that for Brigit to develop …

On the plus side there is a great feeling of freedom showering in the garden, looking down the valley – I'm not sure if it's legal but we still have some devotees.

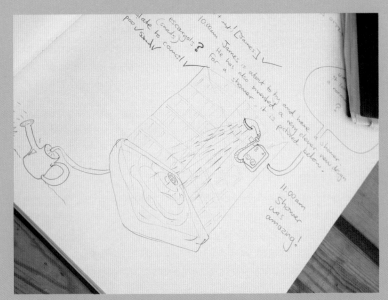

Switching electricity tariff

Even as we were completely preoccupied with plugging leaks and unplugging loos (a regular problem until I traced a pair of 1970s Y-fronts down a main pipe), we made our first and extremely easy act of sustainability. We switched to a green tariff for our electricity. It all sounds simple, and the act of changing was just a phone call, but it did take some effort to make our decision.

Most electricity suppliers rely on fossil fuels, using limited resources. However, it is possible to buy only sustainable, non-fossil-fuel electricity. Some 'providers' don't provide 100 per cent green tariff electricity, and it's not always made explicit in the advertisements. The company making the biggest investment in producing more sustainable power was Ecotricity (www.ecotricity.co.uk), but even with them you are getting only (at the time of writing) 27 per cent green electricity. That said, they are using all their subscribers to get a mortgage to allow them to build more wind turbines. I asked why they provided only 27 per cent green tariff. I liked their logic, which I have paraphrased here: 'Buying already existing green power is not reducing carbon emissions. It is only supporting the *status quo*. It is the investment in new sustainable power and a relative increase in the sustainable sector that helps reduce carbon dioxide emissions. Ecotricity has committed to building more sustainable generators so that the proportion of green electricity increases.'

Here's the bad bit: it costs more – so use less! Remember 'Reduce'. The investment in low-energy light bulbs will offset your costs. Even better, turn lights, etc., off when not in use. Sustainable energy makes sense. I have saved the more controversial comments about coal versus nuclear versus wind versus water for later, when I've softened you up.

41

There is a limited amount of sustainable electricity being made in the UK – stop reading and change your electricity supplier!

The garden – mulching is good

Because we had started our growing in Malvern, planting out was urgent if we were to get any productivity in our first season. Brigit had considered various planting alternatives but, having attended the permaculture course, she had decided that we should follow permaculture processes as much as possible. This involves getting to know your surroundings and slowly planning what will work and allowing it to evolve over time. I thoroughly subscribe to the core principles, but was pleased that we launched into some basic market gardening straight away, which allowed us to get salad and vegetables on the table quickly. We ignored the very sensible and proven advice that states you should live in a garden for a year before you touch it. It was all my fault. I must epitomize the old saying: 'Patience is a virtue, possess it if you can; it is seldom in a woman and never in a man.' The first year worked out and I don't think we did anything irreversible but we are now in a position to take stock and fine-tune.

Masses of slave labour was available, so Brigit and Anda led the way and we blitzed the lower vegetable garden. The vegetable beds were laid out using a mulching/no-till technique – no digging sounds great to anyone who usually gets nominated to do the manual work. If you ask James,

Charlotte or any of the students, all are agreed – mulching is good. Quite simply, cardboard was laid on the cut 'grass'. Scythes and strimmers had been used to take down the top 12 inches of growth and then the lawn mower was put over the top to ensure there was not much greenery left – wherever possible we try not to use petrol-driven or power tools and do things by hand, but if time is short, or we have lots to do, common sense prevails.

The cardboard was soaked to aid decomposition and then compost was laid on top of it all. We are lucky enough to be less

PERMACULTURE: AN INTRODUCTION

Permaculture means different things to different people but at its very heart is the idea of producing a natural, sustainable ecosystem within your buildings, gardens and land. You create a web of beneficial relationships, so that everything you plant is productive without needing the level of input associated with conventional agriculture. For example, a wood containing trees, shrubs, herbaceous plants and climbers needs no tending and relies only upon sunshine and rain, yet it produces more biomass than an equivalent area of wheat, which needs to be ploughed, harrowed, sowed, fertilized, weeded and given pest- and disease-treatment. The trick for us was to create a sustainable ecosystem that maximized the relationship between plants and physical features and also met our needs.

than 2 miles from the Eden Project and Brigit chanced her arm by ringing up and asking if they had any spare compost. After she explained our quest to be green, they said we could help ourselves to a huge pile they had tucked away. We took them at their word and drove the car and trailer through

all the tourists to collect several loads of truly exotic compost. When the beds were prepared and neatly surrounded by brick and stone (both of which we had plenty of), planting commenced. All our nurtured little plants from Worcestershire appeared to love their new home.

43

Living together

All of us – family, friends and the BBC crew – lived in and around the house as we worked. As I have already mentioned, the BBC found some really great people to work with us and it's worth noting right from the start that they stayed with us throughout. We decided that it would be completely against the principles that we wanted to follow for the crew to live in a hotel and come and visit us. It was a good decision – we made some great friends and the crew understood the pressure, the pain and the fun times we went through because they couldn't escape!

Initially we considered various kinds of temporary accommodation (see page 52), but we soon found that we had provided all the mod cons required for a civilized life. Some members of our party had other plans, though. James was particularly keen to live in a tipi, Brigit liked the idea of a yurt and Charlotte would have liked a room with an en-suite bathroom. I'm with Charlotte, but after a lengthy discussion it was agreed that hotels/B&Bs/rented accommodation and commuting were not exactly the green option (and I was very much against the cost factor too). New House Farm is a big old house and even with a lot of work going on there is always somewhere for people to put a bed.

That said, there were several occasions where we had to put up tents as overspill accommodation. We managed to peak at 22 people staying at once – which means a serious number of cups of tea to keep everyone happy. As if we didn't have enough people already, the family are always collecting extra 'helpers' as they travel around. It all started with James meeting a Brazilian musician on the train on the way back from his holiday and bringing him home to stay – they turned up just after midnight on Charlotte's birthday and he started to sing for his supper, breakfast, lunch. Not to be outdone, Brigit met a lovely couple in a lay-by in the Luxulyan valley and discovered that they were interested in medicinal and healing plants – Janet and Peter have since been to stay several times and they have catalogued all the plants on our plot. It's funny how you make friends.

Feeding the masses

Catering was a full-time job in itself. It is fair to say that no two people in the house agreed on what made a good meal. We had omnivores, vegetarians, vegans, carnivores and even some people who thought they were allergic to vitamins and vegetables. So we decided to rotate all our guests through a 'kitchen day'. The concept was easy. Every once in a while you are responsible for a day in the kitchen. Breakfast is always smash and grab, so apart from lunch and supper (and tea and coffee for

anyone who happened to be around), there was theoretically time to chill out or bake or generally do undirected work. We had very different levels of culinary experience. It was an interesting experiment. Meals varied from home-made pizza to tempura-battered courgette flowers stuffed with anchovies. Who could forget Victoria's frittata, or Ant's spring rolls, or Jim's bacon sandwiches? It was great but the budget couldn't always stretch to it and we never quite knew when a meal would be ready. One thing was guaranteed: every pan in the house would need washing by the end.

A bit from Brigit about buying healthy food

It's taken us a few years to switch gradually from supermarket shopping to buying all of our food from local, natural or organic sources. The main reason that we made the change so slowly and over such a long period of time, is that we were put off initially by the cost. I now know that there are cheaper ways of becoming 'green' eaters and, actually, we're not much more out of pocket after all.

We started, as many people do, by buying organic vegetables from the supermarket. The downsides here are that not all supermarkets carry a good range of organic vegetables; they come with lots of packaging; and they are often shipped or flown in from as far away as Australia. In fact, they are hardly ever grown in the UK. Ironically, whilst we were eating fruit and vegetables that made us healthier than those pumped full of pesticides and other nasty chemicals, we were probably doing more harm than good to the planet! So we now buy anything that we don't grow ourselves from farmers' markets and the local grocer, and we have even joined an organic box scheme. In the early days we bought

organically certified meat from the super-market, but we have now switched to buying free-range and local, traditionally reared meat from our butcher. I still look for organic meat wherever possible and know I can always find this at our local farmers' market at Lostwithiel.

When you start buying organic fruit and vegetables, you will notice that the produce does not last as long as non-organic. This is quite simply because the non-organic produce is pumped and sprayed with chemicals and preservatives, giving it an unnatural shelf life. Persevere! It's worth it!

We were then ready to make an even bigger leap and buy almost all of our dried and tinned food from organic sources, as well as making sure that our coffee, tea and chocolate are all fairly traded. We've cut down on the increased expense here by buying direct from a whole-food co-operative which delivers our food in bulk around once a month. Obviously we are able to justify buying in bulk because we are always feeding at least five thousand people (well … almost!).

I'm probably tempting fate by saying this, but should point out that since our food has been completely chemical free, we have also been cough- and cold-free … and that's saved us a small fortune in prescrip-tions and tissues! Before the picture I paint starts to sound too much like life on a health farm, I should add that we think it

would be rude to live in Cornwall and not buy the occasional Cornish pasty or ten. Also, whenever anyone visits us from Malvern, they always bring down a batch of the most delicious scones ever in the whole wide world, made by Mike at Greenlink. I know, it's a bit like bringing coals to Newcastle, but they really do have to be tasted (warmed, with jam and with or without clotted cream) to be believed. Overall however, (and I include the Cornish pasties and Mike's scones here) our eating habits really have become more healthy for both ourselves and for the planet.

If you want to find out what really goes into the food on your plate, you could read *Not on the Label* by Felicity Lawrence. I have learned more about what we buy to eat from this book than I have from any other source, and having read it am now a 'supermarket avoider' wherever possible. Not for the faint-hearted or people who want to wear blinkers!

We had omnivores, vegetarians, vegans, carnivores and even some people who thought they were allergic to vitamins and vegetables

New House Farm recipe

In order to cater to everyone's eating requirements and be as supermarket-free as possible, we have a wealth of recipes for wholesome delicious food that we dish up regularly at New House Farm. Omnivores eat everything; the trick is to make vegan grub appealing so everyone thinks it's a treat. Whether our meals are meat or veggie, we always try to think about where we source ingredients and how we cook them. Here is an example of the sort of food we take for granted with some eco-friendly comments to think about along the way.

Roasted squash soup with chilli and garlic

This is probably our favourite soup. It's filling, tasty and very good for you. Everyone likes fresh bread with soup, but we try to serve alternative carbohydrates some times. We regularly have brown rice and, ever since Brigit was training for the London Marathon, we have had Quinoa *(kee-nwa)* in the house. There are always claims about how healthy foods are – especially if they are unusual. However, weight for weight, Quinoa is richer in protein than meat and contains all eight amino acids! As if that isn't good enough, it has a soft, nutty flavour. You simmer it in water for about 20 minutes, drain and add a dollop to your soup.

GREEN COMMENT: *Pressure cookers save energy and money – brown rice takes about 6 minutes in mine. However, I reckon there are some things that should not be cooked in a pressure cooker. Quinoa, for example, goes to a pretty awful mush – save yourself some hassle and don't try it!*

Ingredients
Any type of squash (de-seeded, skin on, cut into big lumps)
A couple of fresh chillies
4–5 cloves of garlic – its good for you!
Olive oil to drizzle over
Stock – chicken is my favourite, but it's not exactly vegetarian!

GREEN COMMENT: *All the ingredients are homegrown on our plot – the ultimate in food mile saving. The squashes last for months so we can make this all the way through to the spring.*

Method
* Oven on at 180C/Gas Mark 4
* Put the lumps of squash on a oven tray
* Pop on the chilli and garlic
* Drizzle lots of olive oil on top
* Cook in oven for about an hour
* Scrape into a big enough pan. (Make sure you get off all the olive oil and gooey bits)
* Add hot water and stock
* Whizz it with one of those blender things
* Serve straightaway or heat up later

GREEN COMMENT: *It may sound obvious, but anything you can cook a lot of in one go will save energy and money. Whenever you turn the oven on, cook oodles of things at the same time – lots of energy goes into heating the oven and there will always be spare shelves for your flapjacks or roast vegetables. This soup is quick to cook, can lasts for several weeks and can fill a whole camera crew. It freezes well, too.*

Variations

* Could use dry chilli
* More stock for thinner soup; less stock for thicker (we go for thick)
* If you use organic stock be prepared to add more salt
* Garnish or serve with crème fresh
* Butternut squash is a real winner

Temporary homes

Temporary accommodation can vary from something very Heath Robinson that costs a few pennies to something rather grand. There is an amazing amount being made and used in the UK. It all looks rather 'alternative' but that's only because the mainstream has not yet caught up. These are ideas we seriously considered.

Log cabin
Most are now made from wood that is managed (Forest Stewardship Council or FSC, for example) and they can vary from a glorified shed (with thicker walls and nicer windows but definitely not the place to be caught in in the winter) to a serious construction you could comfortably live in for years. We thought it might be nice to have one in the garden near the patio so that once it was no longer required as accommodation it would make a lovely summerhouse. There are lots of people selling log cabins but expect the price to be in the thousands.

Rammed-earth shelters and earthships
Simply ram old tyres/oildrums/tins/rubbish with earth and dig them into the ground to build three sides of your building. Add a bit of render to make it pretty and – hey presto! – a home fit for the most discerning of hobbits. We have a south-facing slope which I thought would be great for vines or tea, but Brigit reckons we could have a lovely little home from home there. I can't quite get my head around it.

Straw-bale houses
When we researched straw-bale houses we discovered that there are examples that have been standing in the UK for a hundred years. The principles are simple and there are numerous self-build books to guide you. The key thing is a good dry foundation. There is something very appealing about having a building with thick walls that you can throw up yourself in a matter of days and you know will retain heat. I suppose it all harks back to the idea of a barn raising, when the whole community got together and helped build something very quickly. If we ever build our own straw-bale building I have no doubt we can rope in all our friends and neighbours for a weekend, but the bill for the beer, wine and barbecue may drive up the cost significantly. It is also worth noting that apart from the walls, everything else in a straw-bale house is conventional – roof, windows, doors, foundations, render, plumbing, and so on. To get a feel for what is possible, check out www.strawbale-building.co.uk.

Tipi
These look exactly like you expect Native Americans to live in! James would have loved one. For me they lack usable head-

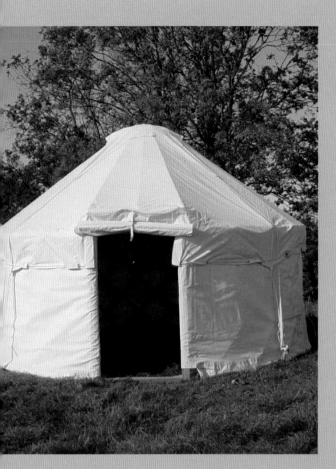

capability. It can be impressive, with an all-natural steam-bent wooden frame that can last 15 years, and a flame-, water- and rot-proof canvas cover that can survive daily living for more than five years and perhaps up to ten. But yurts are not cheap:

16-foot yurt
Frame and canvas: £2,805
Groundsheet: £150
Windows: £90
Wooden doors: £200
Total: £3,245

Optional: wooden floor £500; woodburner from £150

room, and heating is a bit of an issue. There are several suppliers in the UK but new tipis are pretty expensive for their size – about £800 for one with a 12-foot diameter. They are built to last and can be spotted from a distance. (Check out www.tipi.co.uk.)

Yurt

This traditional Mongolian or Turkish tent (also called a *ger*) is built to survive harsh desert winters and, like the tipi, is another design classic. We will either build one or buy one for Brigit very soon. I don't think she will move into it full time, but as a quiet, contemplative space, it would be awesome. The yurt can have proper doors and windows and is intended to have a stove in the middle, so it has year-round

The above Yurt is from Paul King at Woodland Yurts (www.woodlandyurts.co.uk). You could also contact Derick Nelson at Archangel Yurts (01749 890457), where you also get your own 'angel' motif, giving you archangel protection!

General points

There is even a market in great ideas on heating your little green home – sawdust burners are small and cheap and you can usually find a free source of sawdust. However, they do lack the flames that keep you mesmerized for hours. Planning may be an issue as some of the more substantial buildings may not be allowed under General Permitted Development Orders.

Reducing Waste

You could be spending about £470 – a sixth of your annual food budget – on the packaging! That's enough to buy lots of very tasty goodies, loose or in paper bags

There's no such place as 'away'

It didn't matter how we looked at it, we knew we would have a problem meeting our aim for zero waste. To be specific, we were aiming for nothing to be put into landfill. That meant we had to find a home for any rubbish that we managed to accumulate at New House Farm. Brigit was always on hand to remind us that when we throw things away, there's no such place ... so we had to tackle every item that was no longer of use.

Packaging

Nearly a quarter of your weekly bin load is made up of packaging. When you study the packaging you are bombarded with, at first you are struck by how much effort goes into designing, making and using it. There are a lot of people and machines working to safeguard our purchases and it is as if all the packagers in the world are making it their personal quest to help us feel extra loved by giving us more than we could have imagined was necessary. As you begin to come to terms with the startling variety of packaging, you start to get a bit annoyed – well, I did anyway! – at the sheer waste of materials, time and money. Having convinced us to buy the perfectly shaped/coloured piece of fruit or vegetable, or the latest must-have electronic gismo, they now have to present it lovingly wrapped just to show how valuable it is.

While we are on one of my favourite subjects, food, I have a much more fundamental problem with the packaging itself than the waste it creates. I want to be allowed to touch and smell and generally get intimate with fresh produce before selecting what meets my needs exactly. Packaging is a barrier that can hide the truth – I don't like it. Why do I have to have, say, the four apples that someone else has put together? Given a chance, we should all be allowed the very best from each box. Sometimes, when you see a lot of vegetables and fruit nicely laid out to allow you freedom of choice, it is the supermarket lulling you into a false sense of security. Just as you think you are about to have the ultimate purchasing experience, you are thwarted by the packaging police – with the exception of paper bags for mushrooms, all you can find is plastic in the form of those very flimsy bags that aren't quite strong enough for all you wanted to buy, so you have to have two … and the battle goes on! I prefer to shop in a greengrocer where you can put all the produce into one basket and they seem capable of weighing like with like and then decanting them straight into your bag. (They must be on a different training programme.) It goes without saying 'your bag' must be a decent reusable one.

Whenever we have to buy from a supermarket – and there are times when we miss being able to – I now enjoy not taking any plastic bags and watching the vegetables scatter as they go down the little moving belt. Simple things amuse simple minds.

Even with a determined attitude and a thorough understanding of why you don't want packaging, some will find its way to your home. It sneaks in via guests, internet deliveries and all manner of obscure routes. The bottom line is that packaging is only a subset of all the materials that get into your house and, when they have ceased to serve the function they were made for, must be dealt with.

What to do with it all

Organic materials

There are oodles of things that can be done with organic materials. Egg boxes allow us to share our awesome eggs and often return to be filled again. We have stickers to allow us to seal and resend the envelopes we keep getting with free offers, credit cards, etc. And what doesn't go in the recycling bag is composted or used for mulching.

Soft plastics

We keep some plastic bags for unspeakable, smelly, yucky things that occasionally turn up as we have lots of different animals. I'm sure such organic material could be composted, but I'm afraid we like to make it disappear as soon as possible. We collect any soft, clean plastic and, when we have lots, use it to fill bedding covers for the animals.

Bottles, tins, cans and jars

We reuse were possible. We love home-made chutneys and preserves, so jars always find a home. If like us you drink a lot of wine you can collect enough empties to make a heating system for your greenhouse and you don't have to feel guilty (see page 80). If you run out of ideas for your hoards of glassware, it can be recycled, so there is no need for any waste. Include lids when you recycle, and put natural corks in the compost. We also recycle tins and cans.

Foil-lined drinks cartons

Hard to recycle so we don't buy them.

Textiles/clothes

These can go to charity shops or car-boot sales for reuse or if necessary they can go into recycling.

Organizing the recycling

It all sounds very easy but it does take some organizing. We had a fairly rudimentary system when we lived in Worcestershire. We had stacker plastic boxes that had the label from a recycling bag cut out and stuck on the front. It should have worked. It taught everyone to separate compost, paper, plastic, textiles and tins and aluminium cans, but somehow the boxes' position in the utility area and the fact that they were relatively small – because the area they were stored in was quite small – meant that we were always struggling to stop it looking a mess. We tried to keep on top of it but I would like to bet that some things made their way to the bin, not the recycling.

When we arrived in Cornwall we all knew that we were going to recycle our little hearts out. Unfortunately the array of coloured bags (a completely different colour-coding system from the one in Malvern!) languished in the rear yard area as we were running around trying to fix the plumbing, electricity and minor problems like holes in the roof. I'm not making excuses, but it was chaotic. Nothing untoward was sent to landfill and we did manage to get most of the recyclable waste into the right bags on the days that the chaps from our local council, Restormel came to collect, but that was usually achieved on the morning the lorry turned up and involved six or eight people frantically stamping on plastic bottles, tin cans, boxes and anything that had air in it and stuffing it into the sacks. It was not a pretty sight. One day, after about six weeks, enough was enough and the 'New House Farm Recycling System' was born.

Never mind the packaging, every year each of us can throw away as much as £424 worth of food we don't eat or have allowed to go past its sell-by date.
Reduce: buy only what you need.
Plan your meals or get some chickens or pigs!

'The New House Farm
Recycling System'

Anything for the animals is kept in a stainless-steel pot in the kitchen and taken out twice a day when they are fed.

Any composting material is collected in a little compost bin in the kitchen. The lid with the charcoal odour-eater disappeared very quickly and is bound to turn up in one of our large compost bins along with all the missing teaspoons that were probably com-posted with the teabags. We found we didn't need a lid as it doesn't get smelly – the sheer volume of vegetables we are eating means the bin gets emptied regularly and I rinse it in the stream on the way back to the house. No doubt we will eventually have exotic plants downstream of us as the odd rogue seed that fails to go into the composter will manages to germinate and flourish.

There are bins outside the kitchen door marked up for the bags Restormel collects. We also have a couple of specialist boxes: one for spent batteries (we had to use them for the radio microphones when we were filming), and one for metal from the house renovation (mainly copper and lead). Local authorities are committed *only* to providing every household in England with a separate collection of at least two types of recyclable materials by 2010. I am very happy that Restormel are already functioning above and beyond the call of duty.

The bin in the kitchen is for landfill and is limited to plastics the council cannot recycle. It doesn't mean they can't be recycled, just that there is not yet a market for them or a facility available to do the recycling.

Before I leave our simple system it is worth saying that the system has been well worth the time taken to organize, has saved lots of pain and is a reminder to recycle every time we go past it. The family, and indeed most of the people who have stayed with us during the project, would never dream of simply throwing away a teabag or tin or jar or an envelope or anything recyclable. Once you start recycling, it's hard to stop. It makes sense.

Local council collection

✱ Councils all over the country have set up systems so all you have to do is leave the recyclable items outside on the right day.

✱ A good tip is to look out for the well-organized neighbours! As soon as I see mine putting their recyclable materials out, I know what day it is.

✱ Many councils will collect paper, glass, tins, aluminium cans and plastic bottles (tops off). They may also collect cardboard, preferable flattened and in a box.

✱ Check with your local council beforehand and as they should have information readily available.

The recycling process

First of all, there are some symbols to get to know such as the recyclable aluminium, steel and glass symbols. You can have a look at the full range of symbols at: www.biffa.co.uk/getrecycling/symbols.

The processing plant

During the making of the television series I was fortunate enough to spend a day at the local recycling centre. I started by following our recycling bags to the processing area. That was an eye-opener! The area was used both by the fleet of recycling lorries, with compartments for categories of recycled waste, and by the conventional bin lorries, full of black bin liners and miscellaneous refuse. Once the recycling lorry arrived on site, our bags were part of an extremely well-choreographed division into lots of large skips. A couple of pirouettes later and the lorry had successfully separated its load for onward distribution. In the time it took for this to happen, a couple of bigger lorries had come and emptied their guts into the skips to be taken to landfill. It was very saddening to see the bags splitting open and all the bottles, tins, food, clothes and so on mingling in preparation for joint burial. Once everything has been merged like this, there is no way we can expect anyone to sort through the mess.

Recycling has to be done as far forward as possible: namely in the home. It's such a pity that a very large number of people cannot be bothered. I reckon there is a case for the UK government to follow the example of other European countries, such as Denmark, the Netherlands and Sweden, and introduce landfill bans for those materials that can be recycled or composted and where an infrastructure is in place or can be developed within five years. I particularly like the idea of local authorities being given the power to introduce variable charging, making households pay according to the amount of unrecycled waste they produce. Twenty years ago, when I was serving in the army on mainland Europe, we were paying for each (little) bin we had filled. This would provide a real incentive for households to reduce, reuse and recycle waste.

The recycling centre

From our local processing plant, trucks filled with similar types of waste headed for the recycling centre in Bodmin. It was large, modern, smart, clean (yes, clean!) and very well run. The bags were delivered to different areas and dealt with in various ways. One of the least pleasant jobs was the opening of the plastic bags. In our local authority area, we are asked to fill one of our bags with a mixture of plastic bottles, tins and aluminium cans. It didn't make sense to me until I saw what happens next.

The chaps who have opened the bags feed the contents on to a conveyor belt that takes the mixture up to a high-level work-space. Plastic bags are separated from the main cash crop by hand and the first of the clever machines does its bit. A rotating belt with a very powerful electromagnet sucks the cans about a foot off the conveyor belt and sends them into a huge cage. Then the main belt passes an area where an electro-magnetic vortex flings the aluminium cans into yet another cage. The plastic bottles just wend their merry way to a huge pile back at floor level. They are in turn fed up a sort of Archimedean screw on to yet another conveyor belt to be sorted by hand to extract anything that is not PDE, HTPE or PVC (see page 64).

In yet another area the paper and card were passing along a conveyor and being sorted by hand. The waste was then palletized and put on to a very clever articulated lorry that appeared to move the pallet up inside the container by itself. Those of you who have seen this will, I know, have been as impressed as I was. For those who don't have a clue what I'm on about – wait to be amazed when you see a 40-foot container load itself!

At one end of the very large hangar, bags of waste were coming in, and at the other end valuable products were coming out. In fact, the pallets of crushed aluminium cans are so valuable they keep them inside under lock and key at night-time. I was very impressed by what I saw. It is worth going to your local recycling centre if you get a chance. And it makes you very aware of the importance of doing these things:

Thinking ahead

✳ Wash out your cans and plastic bottles – it saves a lot of grief for the chaps who open the bags.

✳ Squash your bottles and cans – it takes only a fraction of the number of lorries to transport them.

✳ Recycling a ton of material at the centre costs the local authority the same amount as it would to put an identical ton in land-fill. It is a myth that recycling is expensive.

DID YOU KNOW? Every year, an estimated 17 billion plastic bags are given away by supermarkets. This is equivalent to over 290 bags for every person in the UK.

Recycling plastics

If you wish to be ostrich-like and shove your head in the ground, I'd just like to remind you that your bottom is in the air and very vulnerable! It's better to face up to the fact that we need to stop using natural resources for energy at the rate we are.

The Strawbridge family definitely took to recycling, but we did have problems in deciding what to do with the myriad of plastics that still managed to sneak into our house. So here is something else to think about: all types of plastic are recyclable.

An example – making carrier bags from recycled rather than virgin polythene has several environmental benefits. It reduces energy consumption by two thirds, reduces water usage by 90 per cent and saves 1.8 tonnes of oil per tonne of recycled polythene produced. Avoiding virgin polythene also reduces the amount of sulphur dioxide, nitrous oxide and carbon dioxide produced.

Different kinds of plastic

For the uninitiated – there are about 50 different groups of plastics, with hundreds of different varieties. To make sorting and thus

PLASTICS AND THEIR MOST COMMON USES

PET	PET	Polyethylene terephthalate – fizzy-drink bottles and oven-ready meal trays
HDPE	HDPE	High-density polyethylene – bottles for milk and washing-up liquids
PVC	PVC	Polyvinyl chloride – food trays, cling film, bottles for squash, mineral water and shampoo
LDPE	LDPE	Low-density polyethylene – carrier bags and bin liners
PP	PP	Polypropylene – margarine tubs, microwaveable meal trays
PS	PS	Polystyrene – yoghurt pots, foam meat or fish trays, hamburger boxes and egg cartons, vending cups, plastic cutlery, protective packaging for electronic goods and toys
	OTHER	Plastics that do not fall into the above categories, such as melamine, which is often used in plastic plates and cups

recycling easier, there is a marking code to help consumers identify the main types. You have to look quite closely at your yoghurt pot to find out what it is! As yet, it is not financially viable to recycle some types of plastic. In our council area they are concentrating on PET and HDPE and a little PVC, and they allow the supermarkets to collect the LDPE.

Purists may wish to tackle the problem of plastic waste by stopping its production. But we have to face facts: it is very useful stuff. Love it or loathe it, we still need to recycle as much of it as we can.

Plastic plus points

* It is versatile and can be tailored to meet very specific technical needs.
* It is more lightweight than competing materials, reducing fuel consumption during transport.
* It is durable.
* It is resistant to chemicals, water and impact.
* It has excellent thermal and electrical insulation properties.
* It is relatively inexpensive to produce.
* It can even be said to have good safety and hygiene properties for food packaging.

> **It should not be that recycled goods are a novelty asking for a sympathy vote. It is important that they are marketed on a level playing field with mainstream products**

What happens next?

As a family, we are committed to recycling, but recycling doesn't make any sense unless we buy or use things made from recycled materials. In fact, recycling hasn't even taken place until a waste product is made into something new. We need to encourage the recycling industry by not only supplying the raw materials, but by creating demand for their products too. I think we will know that recycling is a success when no recycled product bears a label saying it comes from recycled material.

Drinking glasses made from bottles

I was a little sceptical when I went to visit Glen at Green Glass UK. His glasses are very definitely recycled bottles – they don't profess to be anything else. In fact, the original bottle shapes are very easy to see. That said, I liked their quality and utility. Glen got the idea after visiting South Africa, where lots of people are so poor that they have no choice but to recycle wherever possible. His company was started in his garden shed five years ago, with him doing the whole process. In a nutshell, bottles are collected from local pubs and restaurants for a small fee. He even collects bottles in special bins from hotels, holiday parks as well as visitor attractions. They are stored until there is enough of the same type to have a good run through the factory.

The very first thing to happen is the bottles have their bottoms cut off. This is very clever! The bottom is scored, heated and then cooled quickly using compressed air. The resulting thermal shock causes the bottom to break clean off. The edges are ground and melted to leave a nice smooth finish. The bottoms are then stuck on the tops, and the glass is made as tough as possible by passing it through a long oven with a conveyor belt to anneal the glass and ensure that all stress is removed. The glasses take about two hours to go through the oven, which is kept at different temperatures along its length. Finally they are printed with a funky design.

Glen is setting a cracking example because his product can compete with others made from conventional resources. Not only are the bottles recycled, but because the process doesn't involve melting down the glass, it saves 90 per cent of the energy normally used in recycling glass, too. The business is strong and such is the company's demand for bulk deliveries of bottles, it is also creating a market for sorted, recycled, whole bottles by buying bottles from people who have set up as professional recyclers. These recyclers separate out bottles from kerbside and pub collections and sell them on. Glen will also buy bottles in bulk, so any companies out there with waste bottles, contact him at www.tradinggreen.co.uk.

The Eden project has competed under self-imposed ethical restrictions and it continues to thrive

The Eden Project approach

We are fortunate enough to live very close to the Eden Project and have visited on several occasions. Their mission – 'to promote the understanding and responsible management of the vital relationship between plants, people and resources, leading to a sustainable future for all' – shows they are a great source of information and inspiration. They are an overt example of sustainability, covering all aspects with their systematic approach.

The Eden Project is a big business. In fact, it's better than that, it's a successful big business. It aims to balance the financial, social and environmental impact of its actions. It would be great if other large British businesses decided to follow suit – who knows what would happen?

When Brigit and I visited Sustainability Director Chris Hines at their Waste Neutral recycling compound, it quickly became apparent that they take their responsibilities very seriously and that they are prepared to invest to achieve waste neutrality. It's also worth mentioning that the Eden Trust has deliberately set out to operate in the commercial arena. They believe that only by demonstrating that ethical commerce is viable can they effect real change in the global businesses they would like to influence. With more than 200 suppliers locally, their first year of trading put around £150 million of additional revenue into the local economy. They are huge players and have laid the foundations for concerted strategic action among their suppliers to deliver social and environmental benefits. They are attacking the waste problem in several ways.

First, waste on the site. There are so many visitors and they produce so much waste that recycling is a massive operation. The largest on-site initiative has been the creation of a new recycling compound,

SOME THINGS ARE HARD TO ACHIEVE
It is a prerequisite of its planning consent that 20 per cent of visitors to the Eden Project come by means other than a car. The Project has a phenomenal number of coach bookings and in the fullness of time would like its own rail link. However, no one should be in any doubt that the transport of choice for most people – environmentalists too – is still the car.

receiving Eden's waste for recycling but also containing public areas such as a workshop, exhibition and viewing space called the Alchemy Centre. Eden has worked with a bin manufacturer to design the 'Ultimate Recycling Station', streamed wastebins designed to encourage and make it easy for visitors to recycle their waste. Recycling needn't be a complicated, time-consuming process, and Eden hopes to demonstrate this so that visitors will take the message home and perhaps step their recycling efforts up a notch or two. Eden aims to buy in the same amount of product made from recyclates as it sends out for recycling. This has the advantage of stimulating the market for recycled goods.

Second, they ensure as little waste as possible in their supply chain. Waste-neutral activity with their suppliers and producers takes place in two main areas – reducing their waste and prioritizing the use of recycled materials in their purchasing policies, and encouraging the development of new products made from recycled material.

Finally, and probably most important, they have undertaken to inform and inspire people to take action to reduce their own waste production at home and at work.

THE WEEE MAN

We are gradually getting to know our way round the Eden Project. Every time we visit we are struck by the WEEE man. WEEE stands for 'Waste Electrical and Electronic Equipment' and the giant sculpture represents the total amount of electronic waste an average person in the UK will discard in a lifetime. He has computer mice for teeth, an old washing machine for a spine and his neck is made out of vacuum-cleaner tubes. He is a 3-tonne, man-made monster, and when you look at him you recognize lots of things you have owned, used and discarded, and not always because they were beyond repair. Actually, it is quite uncomfortable looking at him.

Our way forward

Having tried to produce no waste, we have failed. That said, as a family of four we probably produce less than a carrier bagful of landfill per week. We will try to reduce that, but it's proving hard. Some things that are worth doing:

✷ Keep a bag in the car for when you pop into the shops. And whatever you do, don't weaken and individually bag up your goodies once in the shop. If you must go to a supermarket, let your vegetables roll their way down the conveyor belt and defy anyone to challenge you.

✷ Make sure you buy to last. Get something that can be repaired. I like well-made equipment. It will cost more, but a reliable make will still be going 10, 15, 20, even 25 years later. Think about buying second-hand if you cannot afford something new.

✷ Cancel your junk mail.

✷ Organize your recycling area.

✷ Visit the greengrocer.

✷ Look at the amount of packaging before you buy something. It doesn't count if you ask to leave the box and/or plastic bits in the shop – they will probably put it out for landfill. Remember the ostrich with the vulnerable bottom?

A bit from Brigit on household cleaning and chemical waste

There seemed to be little point in making such big changes to our waste management without going the whole hog and also looking at our household cleaning 'stuff'. To look at the way the army of bottles and sprays we used to have under our kitchen sink had been marketed, you would be forgiven for reaching the conclusion that our home was a war zone! Most of what we'd been using on our baths, floors, clothes and surfaces for the last decade or so would be capable of killing a living organism and it's a bit horrifying that they get washed down the drain, polluting the earth, rivers, seas or atmosphere. Remember, there is no such thing as 'away'!

I believe there is a balance to be achieved and the occasional strong chemical in the house is hard to avoid, particularly if you have small children or animals and need to keep the place hygenic. For everyday use, however, there are many products on the market which are much gentler on the environment. We have been using Ecover (08451 302230; www.ecover.com) and Clearspring (01617 642555; www. faithinnature.co.uk) products for some time now, and I am completely satisfied with their performance. Another plus with these products is that you can often refill the bottles at your local healthfood shop, which saves you having to recycle them. Reusing is better than recycling!

If you want to know exactly what's in your cleaning products you could always try making them at home. You can go a long way with vinegar, lemon juice, bicarbonate of soda, borax and soap flakes. Try the recipes in *Organic at Home* published by Murdoch Books – they couldn't be simpler and they really work!

To look at the way the army of bottles and sprays we used to have under our kitchen sink had been marketed, you would be forgiven for reaching the conclusion that our home was a war zone!

Self-sufficiency

For the Strawbridge family, self-sufficiency is about having our own produce on our table as far as possible, but not about being isolated from the outside world

Our own early harvest

We want to live a 21st-century life. It is not acceptable to us to be slaves to the land and the animals. I have to continue to earn a living and everyone here needs space and time to enjoy life, so we don't want a commitment that would take over. It was with that in mind that we launched our drive for self-sufficiency.

Food is important to me. One look at my rather sub-statial frame will tell you I am someone who likes to eat. Not only that, I like to eat well, eat healthily, eat tasty stuff and eat lots of it (must be my Irish up-brining!). Brigit taught me to cook when we first got married and gradually I started to push her out of the kitchen. When I come home late from work, I relax by preparing a meal and having a glass of wine. When I left the army I was allowed to undergo training to prepare myself for life as a civilian, so I decided to go on a cookery course. I then proceeded to buy a first-class cooker, which we sold with the house in Malvern. When we moved to Cornwall, we discovered that we could not do without such a cooker, and so, despite it needing gas and being huge, our Mercury cooker has pride of place in the kitchen. (I still harbour ambitions to make my own biogas/gasificator, so I may yet be able to get rid of our need for fossil fuel.) As a family, we eat most foods, though Brigit and Charlotte prefer to eat less meat. Our friends and visitors vary from vegans to fast-food junkies, however, so we have had to cater for all tastes.

While we were planning to harvest our own vegetables and waiting for our pigs to grow big and strong, we were reliant upon shopping. We began to question what we were being offered. We wanted to buy organic produce wherever possible, but I have a problem with the organic sections in most supermarkets. They tend to be crap. I like buying

British which means buying seasonal and local produce with outstanding flavour. There is a time and place for every fruit and vegetable – it makes the first one you taste each year all the more special. Generally, growing your own vegetables allows you to look more critically at how they are being produced elsewhere.

With a well-stocked larder of dried goods and using as much locally produced meat and vegetables as possible, we waited to be able to harvest our own crops. It didn't take long and efforts in June were reaping rewards by the beginning of August. In addition to the quick wins of seasonal salads and vegetables, we also started to plant some perennials and plan for the future.

Our plot is truly lovely to look at. We have lots of different zones that can be used in different ways. With such potential it was a matter of enjoying the garden and planting as much as we could in the time available. We were fortunate enough to have Patrick Whitefield, author of *The Earth Care Manual*, visit a couple of times to give advice. If you haven't seen this book, have a flick through it in a bookshop – you are sure to decide you need it. Brigit had attended his course at Raglans Farm and we knew we wanted to embrace the principles of permaculture as far as possible. Autumn saw us committing to some of our zones and planting trees.

Our plot

The house and four old stone barns sit in the middle of the plot at the end of a southwesterly-facing valley.

To the east lies the upper paddock, which was known as the Abbey Orchard and the most likely location for the ruins of the medieval priory. With walls on three sides, this is the home of the greenhouse (on the south-facing wall), has the aqueduct running east–west through the middle and houses various vegetable patches as well as the pigs.

To the south on the hill is the garden, enclosed by a stone wall and full of exotic plants (they aren't much use but people seem to like them) with a lovely patio area near the house. There is no through lane that will eventually grow mushrooms.

To the west down the valley is the lower paddock with its spring, which is completely independent from the stream that runs through the paddock. The pigs started their life here but it was too damp, so we moved them. We also have a potting shed and a pole barn (a frame of wood with telegraph poles, covered in corragated iron) there. The chickens have the run of this area. We have yet to exploit the south-facing bank, but it has great potential.

To the north, by the entrance, is an area sheltered by leylandii. This was our first vegetable patch.

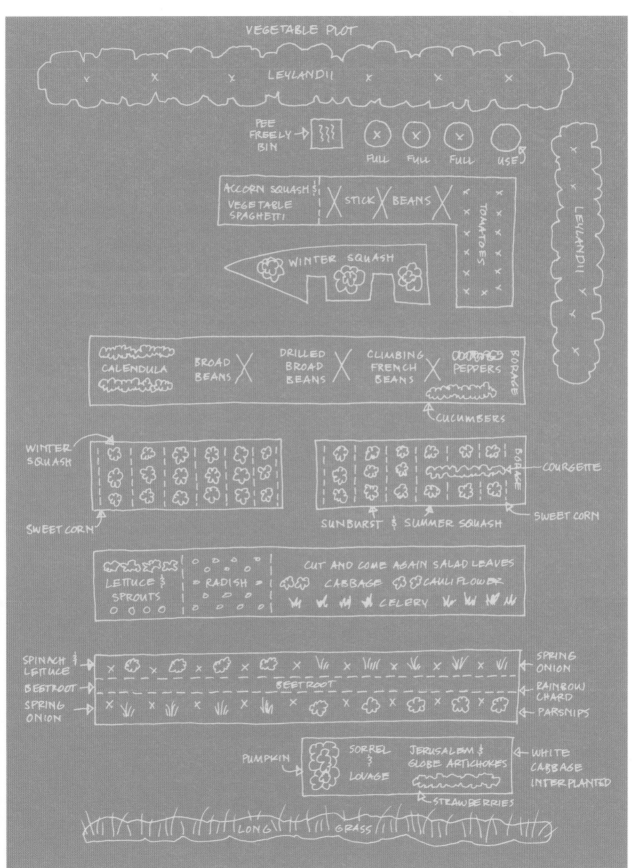

VEGETABLE PLOT

LEYLANDII

LEYLANDII

PEE FREELY BIN →

FULL   FULL   FULL   USE

ACCORN SQUASH &
VEGETABLE SPAGHETTI

STICK BEANS

TOMATOES

WINTER SQUASH

CALENDULA

BROAD BEANS

DRILLED BROAD BEANS

CLIMBING FRENCH BEANS

PEPPERS

BORAGE

CUCUMBERS

WINTER SQUASH

COURGETTE

BORAGE

SWEET CORN

SUNBURST & SUMMER SQUASH

SWEET CORN

LETTUCE & SPROUTS

RADISH

CUT AND COME AGAIN SALAD LEAVES

CABBAGE   CAULIFLOWER

CELERY

SPINACH & LETTUCE →

BEETROOT →

SPRING ONION →

BEETROOT

SPRING ONION

RAINBOW CHARD

PARSNIPS

PUMPKIN

SORREL & LOVAGE

JERUSALEM & GLOBE ARTICHOKES

WHITE CABBAGE INTERPLANTED

STRAWBERRIES

LONG GRASS

The greenhouse

If you are serious about providing your own food, it is essential to try to stretch your growing season. With the south-facing wall in the upper paddock, the obvious answer was a greenhouse that used the warmth the wall would store during the day to reduce the effect of cooling at night. We managed to find a reasonably priced greenhouse which we felt we could attach to our wall. Well done, Richard at Greenhouses UK (www.greenhouses-uk.com)! He believed us when we said we could put up a working greenhouse with one side missing.

The first problem: the wall was in a serious state of disrepair. However, we had a bunch of workers who didn't know the meaning of the word 'impossible'. So much for university education! James, Henry, Ant, Vankey, Nsa, Andy (Jim's brother) and Nick the soundman all pitched in. Having excavated the spot where the greenhouse was due to go, they built a wall to mount it on. The excavation was not trivial; the rocks and debris from the wall were under brambles that could have kept Sleeping Beauty undisturbed for another hundred years. Some brute force with the scythes, some energetic riddling with our specially developed two-man riddle and a large amount of determination later and the wall was visible. The discovery of bones did cause some consternation but they were

very large and it transpired they came from an old pig, so they held up proceedings for only a short time. The building of the wall became an artistic endeavour. All the rocks had to be just so and the chaps even managed to design in some features, such as a special shelf to hold a teacup. All manner of things decorate the mortar, too: a silver frog, crystals, even a triceratops.

Wall built and foundations laid, we had the frame up pretty sharpish (again thanks to Richard – he knew what he was doing!) and even the glass was pretty painless to fit after we had invested in a roofer's square (an invaluable right-angled bit of metal). Actually, pretty painless probably doesn't accurately describe the hours Tot, a friend from Malvern, and I spent trying to get it exactly square so the panes would fit. When I was called away to other things, I'd come back to Tot who would have done a three-dimensional problem-solving calculation along the lines of 'pushing in here will cause this acute angle to become obtuse ...' He was not going to be beaten!

Before we fitted all the glass and let the planting commence, we had one last trick up our sleeves: a very sexy little system for keeping the greenhouse warm. We'd met Steve from Krysteline (www. krysteline.net) when Brigit and I had been visiting a company to talk about eco-paints. He was servicing a machine that imploded glass so that the bits that were left were not sharp.

It's very hard to say exactly how efficient our little system is but, combined with our wall, it has kept our greenhouse frost-free and everything in there looks great!

Much in the way that an opera singer can set up a standing wave in a glass that can cause it to implode, Steve's machines set up similar waves. (I hope Steve and his colleague Stefan will excuse my much-abbreviated explanation.) Steve had hundreds of uses for it, from producing cosmetics to aggregate, but the property that caught my attention was its thermal performance so I managed to convince him and Stefan to come and implode some glass for us. We had a polystyrene-lined hole in the greenhouse with a series of plumbing pipes in it, for air to circulate through. We then filled the hole around the pipes with little bits of glass. We turned the machine on and fed it the contents of our glass recycling bins. The reduction in volume was huge, about 10:1.

With our supply of glass used up, James headed off to the pub and we took all theirs as well. A couple of hours later we had a very efficient heat sink in the floor of our greenhouse.

The whole greenhouse process was no mean feat considering we regularly had frosts, sometimes down to –6°C and we even had 5 inches of snow. So much for a mild Cornish climate. Every time we would have spent £1.99 for a bag of mixed salad leaves at the supermarket, we had our fantastic home-grown leaves to use instead. They continued to grow when we cut the leaves off so I think its fair to say the greenhouse had paid for itself by the end of February: not a bad investment.

How our greenhouse heater works

❋ The hottest air in the greenhouse rises to the apex of the roof. A fan sucks the air from the apex and blows it through the bits of glass under the floor of the greenhouse. All day long, when the greenhouse is at its warmest, the air warms up the glass heat sink.

❋ As the greenhouse cools, the air from the apex continues to blow over the warm glass and is warmed. This warm air is blown into the greenhouse.

❋ The fan comes from a computer. It cost about £1.50 and is built to run on very low power. The box it is in is made from a small offcut of plywood.

❋ The pipes are old bits of wastepipe from the bathrooms.

❋ The battery is charged by a small 11-watt solar panel that cost about £35; the charge controller that stops the battery overcharging cost about £7.

In all, not a bad first year. Seed harvesting was an investment for next season and was very successful. We have masses of our own compost, so we don't have to pay for anything from here on!

Salad leaves	Absolutely masses of exotic leaves. The variety and tastes were excellent. A definite success. We even had some in the greenhouse through the winter.
Celery	The leaves and stems were very strong-flavoured and great in salads. I love cooked green celery. Unfortunately I'm in the minority, but we still grow it.
Beetroot	Baby beetroots roasted with fennel and balsamic: awesome. Good in soup.
Radishes	We think that there is a chance the organic seeds may have mutated! They were positively huge, very hot and gorgeous.
Herbs	We lost a lot of our soft herbs: parsley, coriander, etc. (something ate them when they were young). Cut them off at ground level; they never stood a chance!
Beans	(Green, runner, broad and French) I love the look of beans growing in the garden.
Rainbow chard	Looked good, grew well and we used in salads as well as a vegetable side dish.
Peppers	(Sweet and chilli) Not bad, but we needed more.
Tomatoes	Didn't like it outside. There were only a few nice juicy red ones, but we made some great green tomato chutney.
Courgettes	(And patti pan) Great, prolific and healthy. Cook them whole in butter and garlic.
Cabbage	(White and red) Bloody caterpillars! They had more than their share of our cabbages and everything else leafy. It was not a good year for our Cornish cabbages.
Turnips	A surprising success. They seem to like our soil and were big and bountiful. It's amazing how many ways you can do root stew!
Broccoli	The sprouting broccoli had a really long season and the more we picked, the more it grew. We intend to grow more.
Artichokes	Year one for the globe artichokes, so we didn't expect much, though Brigit made a very impressive table display – beautiful blue colour. The Jerusalem artichokes yielded masses.
Spinach	Great young in salads; great old in soup and as a vegetable. Just great!
Kale	The caterpillars didn't seem to want to eat it, but it was very tasty. We had some of the die-hard junk food eaters admitting it was really tasty.
Leeks	When you grow your own there is even an incentive to eat the green bits.
Swede	Small, but tasty.
Parsnips	They were tiny! They tasted magnificent, though. We need more and bigger next year. Jim loves them and waited months for a couple of meals.
Fennel	They didn't look perfect, but tasted great.
Sweetcorn	The little ears were very tasty, but being planted late they didn't get big enough.
Brussels sprouts	They didn't quite firm up and suffered because of the caterpillars. However, the top leaves were deliciously cooked.
Celeriac	Only one! The rest were eaten when young.
Squashes	Pumpkins, onion, acorn, spaghetti, butternut – they all looked great and kept reall well. Somehow we didn't have as many butternut as expected, but the rest had good yields.
Spring onions	Sound and tasty, lasting well into the winter.

Benefits of composting

● Recycles organic waste.

● It's free.

● Improves the soil and maintains fertility without chemicals.

● Reduces landfill.

● Saves on scarce resources like peat.

Compost: where waste management and gardening meet

It is great having a home for any organic matter we need to dispose of. Compost is the end product of a decomposition process involving all our organic waste, including domestic waste (vegetable peelings, eggshells, cardboard, paper – but not shiny, non-biodegradable packaging – teabags, coffee grinds, withered flowers, even the contents of the vacuum cleaner). It also includes garden waste (weeds, strimmed vegetation, weed and silt from the stream, leftover vegetable bits, grass clippings, manure from the chickens and pigs) and anything else we can find that looks like it should rot nicely.

Compost bins

Right from day one we have been maintaining our compost heaps and we are now at the stage where it is not possible for us to put a teabag or anything else in the landfill bin that could otherwise be decomposed. We inherited about ten very large black plastic compost bins when we moved in and the initial clearing of the vegetable patches just about filled the majority, so Anda set about building some more out of our scrap wood.

There is very little we will not add to our cocktail of goodies: when we visit the beach on the other side of our hill, we always try to collect a bag of seaweed, and we are all used to the sign, 'Please pee freely here'. The chaps do their bit, though there are times of the year when the hedgerow is a little sparse and we have to curtail our activity for fear of scaring the neighbours.

Today, we have composting going on in lots of different areas of the garden and we are using our very wholesome compost for the second generation of plants.

How composting works

One of the best things about composting is that it is a natural process that goes on anyway. As gardeners we try to organize it so it fits in with the management of our outside space.

Composting is logical and once you understand some of the elements you can play tunes to suit yourself. Decomposition relies on bacteria which need air and water to flourish. If you use up the available oxygen, the microbial life (the bacteria) in your compost will perish. To reoxygenate, it is necessary to mix and let air in. Ideally, the moisture content should be about 50 per cent. Less, and the process slows down. More, and the water takes up the space of air and the process slows down.

If you manage to achieve a high enough temperature in your compost, you have the advantage of killing off weed seeds and diseases.

85

Getting rid of cooked leftovers

Rats can be a problem so we don't like putting any cooked food in the compost, though it shouldn't be long before our growing cat posse start earning their keep and getting rid of the rodents. We moved from Malvern with a couple of geriatric cats, Baggins the huge ginger tom, and Megan our three-legged, spherical, grumpy cat. In September reinforcements arrived in the form of the kittens Dolly, Willow and Twig. They are full of beans but appear to share the same brain cell, so heaven knows when they will get on top of the rat and mouse issue – we wait with bated breath.

Instead we use the animals for disposing of cooked waste whenever possible. The only organic matter we put in landfill is leftover pork products and trimmings, but usually not much of that gets left. Brigit and I visited Chris Hines, the Sustainability Director at the Eden Project, to find out what they did. They had just invested a small fortune in a high-temperature composter specifically for cooked waste. Regulations do not permit businesses to keep pigs to 'recycle' their food waste, otherwise I'm sure Chris would have had a herd to be proud of. The Eden Project has thousands of visitors and takes waste management very seriously. Very few institutions would be prepared to commit as much time, effort and money to their recycling systems as it does. What's more, most of its recycling initiatives have not only reduced waste but have also proven to be cost-effective in the long run.

If you don't have the facility to feed leftovers to your pigs or chickens, there is something called a Green Cone, which looks

Bacteria at Work

❁ When the composting process begins, the bacteria that survive at low temperatures of about 13°C/55°F – the psychrophiles – start to digest and alter the state of the material in your compost pile. This action gives off energy and the temperature within the compost starts to rise.

❁ Next come in the real workers, the mesophiles, who like 20–30°C/68–86°F. They are seriously efficient. When they are working, the environment is perfect for worms to help break the material down to nice little bits.

❁ Sometimes, if you get it exactly right – that is, you have a substantial amount of carbon-rich and nitrogen-rich materials, with the right amount of water and air – the temperature can rise even further and allow the thermophiles (40–70°C/104–158°F) to do their thing. This frantic decomposition dies off after a couple of days, but if you mix to allow more air in, it is possible to bump-start it again.

86

really useful and should work anywhere – you don't need a lot of land to maintain it, just a sunny well-drained site. There is no end product, as with a composter, though the nutrients do seep into the surrounding area to increase fertility. There appears to be no fixed price for the Green Cone. The makers sell directly to the councils, who set the price. Our local council, Restormel, are currently doing a promotion on this product; it might be worthwhile contacting your council to ask if they can get you one.

Food rotting in a landfill site is not the same as composting – because of all the other waste, it will not decompose as well or produce as many nutrients. If you have any questions, phone the Green Cone customer helpline on 0800 731 2572 or go to www.greencone.com.

Benefits of the Green Cone

- Easy to set up and maintain.
- Immediate disposal of waste food.
- Clean dustbins.
- Other household waste is clean for recycling.
- Reduces household waste by 20 per cent.
- Less waste means fewer lorries on the road.
- Reduces the need for landfill and large-scale treatment plants.
- Nourishes your garden soil.

We live in a world where meat
is no longer valued ...

More to life than vegetables – our attitude to meat

Meat is expected to be widely available and cheap. When I go to the butcher's, I know what I want and I expect quality. I like tasty, well-cared-for meat, I like to see good butchery, I like good advice about the cut and the cooking, I like a smiling face and, most of all, when I get it home I like everyone around the table thoroughly enjoying good grub – and, what's more, I think I'm very reasonable to expect it all, all the time! Producing meat takes a lot of time and effort, but when done well it is worth celebrating.

As a family we used to eat a fair amount of meat, not every day but most days. Things have been changing over the last couple of years. We decided to spend more but less often. Since moving to New House Farm, we have been eating meat at only three or four meals a week. Having said that, we are blessed with an absolutely brilliant butcher in the village – Charles Harris. He knows his trade and provides a service second to none. It is a pleasure to go and buy meat. There are always lots of people there and the choice is mouth-watering. The standard of meat in the village makes New House Farm an even better place to live – how lucky can one man be?

Getting pigs

New House Farm is a registered smallholding and that meant we had the opportunity to put our money where our mouth is. So we did. We immediately sought advice from David Bailey, the third generation at our local farm suppliers Walter Bailey Par Ltd. He had a fair amount to say and was very keen to establish that we intended to rear the pigs properly and not just play at farming. We passed his scrutiny and he and his father, Andrew, have been a godsend. We went to visit his sows and weaners and, having booked two healthy young gilts, set about getting ready for them. It wasn't that difficult. As always, there is some serious expenditure at the beginning, which means if you only ever rear two pigs they will have cost a fortune.

We prepared an area for them in the lower paddock, which involved cutting down the grass where the electric fence was

ITEM	COST(£)
3 pieces of curved corrugated iron	£60.00
Plyboard and wood for the sty	£45.00
Electric fence wire	£35.00
20 fence poles	£20.00
Charging unit	£94.00
Food trough	£18.00
Water container	£14.00
Organic feed (44lbs, enough for 5 days)	£8.70
Total	£294.70

going to go so that it would not spark to earth. Then, we put in our poles and rigged three strands of electric fence all the way round. The area they had was a reasonable size – 32 x 98 feet – and there was plenty of ground for them to dig over. To ensure they didn't turn over their water container, we dug a hole and sunk the container about 6 inches into the ground, and then placed a rather large rock in it. We made them a wallow by adding lots of water to a muddy bit of the pen – someone even attached a parasol to a pole to provide shade. I drew the line at deckchairs. With all the preparations done, we waited for David to deliver them next morning.

Pigs are intelligent, clean and can be very friendly. They will never foul their sty and are always pleased to see you – though that could be because I'm usually carrying their food. Our weaners were Saddleback crossed with English Lop-eared, commonly called Silverbacks. They had been kept under cover in a large walled pen, so our paddock was very different for them. They were about to taste freedom for the first time. When David arrived, he reversed down to the pen, dropped the trailer tailgate and we ushered them in. As soon as the tailgate was up, we switched on the electric fence. Sure enough they both touched the fence and squealed, but they learnt very quickly and it wasn't long before they were happy trotting round and snuffling up roots. As we knew they were destined for the table, they were not given names, lest people forgot why we had them. I liked Jim's idea of marking their joints with indelible ink as a way of reminding people they were for the pot. When I fed them morning and evening I used to go in the pen with them and give them a scratch. Pigs are big beasts and I reckoned it was better to keep them used to people. It appeared to work.

The pigs' house move

By late October, when the pigs had completely turned over their area in the lower paddock and the rain had made it too damp for my liking, we decided to move them to the upper paddock. Six of us did the move: myself, Brigit, Anda, Jim, Nsa and our friend Natalie White, who had popped down for half-term. The BBC crew played their 'we need to capture this for posterity' card and stayed well out of the way. We had a briefing session, and four pig boards – large bits of thin plywood with hand-holds that allow you to restrict the pig's view and encourage it to go where you want – were issued to Brigit, Jim, Nsa and Natalie. Anda and I were to have bowls of food to encourage the pigs to follow us on a rather circuitous route to the upper paddock. Having given them half rations the day before and no breakfast, I was confident they would be keen to follow. I split my forces into two teams and pre-positioned them ready for the off. We turned off the fence, removed it and started to cajole the pigs. Absolutely nothing

happened. The little buggers would not cross the line where the fence had been – who said pigs were bright? After a good ten minutes of coaxing them, we changed tack – there is a military saying that no plan ever survives contact with the enemy, and the pigs were definitely not friendly forces!

Our four intrepid pig-board carriers formed an open square around the pigs and started to move, nudging our unwilling foe forward. It was going rather well until the squelchy noise that indicated that Natalie's boots were determined to stay behind. I'm glad to report that the troops held firm and after we had moved the monsters about five paces they got the hang of it and took off by themselves. I managed to stay in front of them and within two minutes they had passed the chickens, gone through the yard, round the front of the house, up the bank behind the old barns and across the upper paddock under the aqueduct, and were contentedly munching from their trough. They did not seem at all fazed by their adventure and lived out their lives in luxury with lots of new ground to turn over.

91

It would have been
hypocritical of us to
enjoy meat from the butchers
and not to face where
it had come from

Slaughtering the pigs

I like rearing pigs, I like knowing where our meat comes from and I like eating home-produced pork. Indeed, I'm keen to try to rear a variety of animals for their meat.

Chance would have it that when we met Mike Paull, who was keeping some Highland cattle in the field next to us, we discovered that he had been a slaughterman. When we got to know him better he agreed to helping us kill the pigs, which meant that they would not have to travel to an abattoir. I was over the moon as it meant that we were able to avoid what would have been a relatively stressful experience for them. Mike had suggested that we move them down to one of our outbuildings the day before we were due to kill them. They would then have the opportunity to settle down before he came along to dispatch them.

As we wanted a pig for our planned New Year party we had to kill one a couple of days before we needed it so the flesh would firm up. When it came time for the move we were a bit short-handed. It had been a family Christmas with just the four of us, but Charlotte had gone to Malvern to her friend Jim's 21st birthday party and Brigit was having nothing to do with it, which just left James and me to move them. Actually, it went really well. They followed us down to the outbuilding and obediently went in. I felt really rather guilty and a bit sad that they had been so trusting, but I consoled myself in the knowledge that not only did they not know what was happening, but they seemed really pleased to have some-where else to explore. Mike came around late morning and we were blessed with a lovely crisp December day. There was not one cloud in the sky. Getting down to the details, we had planned the day well in advance and had bought or made all tools we needed.

I moved the first pig to one end of the barn to an area that we had sectioned off to restrict the pig's movement and to make the job easier for Mike. Mike took his time and as the pig was feeding he lined himself up and shot it in the head with a .22 para-bellum round that killed the pig instantly.

BUTCHERING EQUIPMENT

It was important to check we had all the correct equipment:
* Mike's tools of his trade.
* Rope or chain to attach to the pig's leg.
* Winch to raise the dead pig.
* Knife to bleed.
* Tray to catch blood.
* Bins for bristles and unwanted innards.
* Butcher's block to work on, mounted on breeze blocks.
* 'Candlestick' scraping tool.
* Water boilers and watering cans.
* Brushes and brooms.
* Aprons.

I can put my hand on my heart and say our pigs had a great life that was ended without any stress

We attached a rope to one of its rear legs and hooked it on to the hydraulic lift of Mike's tractor and James then hoisted the pig in the air (though dead, it kicked quite a lot at this stage). Mike cut into its throat to bleed it – we caught the blood for black puddings – and we transferred the pig to the butchers block and proceeded to take the bristles off it.

I had fixed up an old boiler Mike had picked up at a farm sale and I had an old Burco, so we had gallons of hot water on tap. The water was poured on to the pig to scald it, enabling you cleanly to scrape the dirt, bristles and the top layer of skin off. (Mike had described the tool he used to use to scrape with, a 'candlestick', but as he hadn't managed to find his I'd made a couple out of 10mm 'all thread' and 4-inch hole cutters with all the teeth ground off. They worked a treat!

Once it had been completely scraped and cleaned Mike's final act was to cut around the pig's nether regions so that when we attached the pig to the hoist again it was much easier to paunch it. When we hung up the now-prepared pig it was just a matter of getting everything ready for the second.

It probably took us just over an hour and a half from segregating the pig to hanging up the carcass, during which time the second pig was either munching at some carrots or sleeping. It appeared completely relaxed. We had a well-earned cup of tea and then did it all again. It sounds a bit odd but I was positively happy at the end of our work. It was not because we had killed the pigs, though it has been an ambition of mine for years to rear and kill my own meat; it was because everything had gone so smoothly. I truly believe the pigs never knew what happened to them.

No wastage

You can eat every part of a pig, and we did. A couple of days after we killed them, Chris came down from the butchers after work and helped me cut up the larger of the pigs. How lucky am I? We spent a couple of hours and at the end of it we had roasting joints, piles of meat (good meat!) for sausages and salamis, slabs of meat to use for bacon and ham joints all neatly arranged and ready to use, freeze or cure. We had bought a stand-up freezer in anticipation (high efficiency obviously!) and we filled it with one pig. When Chris finally managed to get home after a very quick cup of tea, I went back to the out-building and started to dry-cure my bacon. I had decided years ago that I wanted to dry-cure, so I went for a blend of salt, juniper berries and crushed peppercorns. It took less than a week (I visited the cure box twice a day which is more than is necessary) and we had bacon and hams. However, I did put a complete leg into its own cure box for six weeks (with a weight on it) and that is going to be a complete air-dried ham in the next six to nine months – this is definitely not fast food!

While the bacon was curing there was still a lot more to do with our meat. For Christmas 2004 Brigit had given me a first-class sausage-making machine and I was not going to miss out on the opportunity to

make my own sausages from our own pigs. Having cleared the surfaces in the kitchen, we minced our pork and then made pork and apple, pork and leek, and pork and garlic sausages. It was a great couple of hours, we had a kitchen full of helpers and we minced a serious amount of meat. We also made a selection of salamis that spent two months hanging in our back yard maturing. James, Ant, Tom and his girl-friend Laura were not phased by the mincer getting blocked or by trying to feed the casings on to the nozzle, or any aspect of the task. I'm glad to say they were focussed on the objective – sausages and salamis. Our pork and leek sausages used our own leeks, of course, and, by special request, we made some chilli and red wine sausages (they made me do it, honset!). Before you commit to filling your sausages you have to test the seasoning by cooking a bit. The job went down so well that we ended up cooking some of the sausages just as they were coming off the end of the machine.

The fate of one of the smaller pigs was

very extravagant. We spit-roasted it to celebrate the end of our first year in New House Farm. Jim and his mate Andy built the spit roast a couple of days before New Year. We invited friends to come and celebrate with a lot of our local real ale 'Tribute' from St Austell and a very healthy supply of wine. We lit the fire at about ten o'clock in the morning. We had intended to automate the turning of the spit, but time was against us so we manually turned it every 10–15 minutes. It was a long day and it was great to see reinforcements arrive in the form of James Cooper, an old mate, who threw himself into pig duty. I don't think we had a beer before about midday, but it was thirsty work, and we didn't declare the roast ready to eat until some time after nine in the evening, by which time the chefs all smelt a bit kippery – it was smoky out there. After all that effort I'm happy to say that the pork was very tasty indeed. We had set up a marquee and live music was courtesy of Charlotte and her friends Beth and Bobby, from Newcastle University, Geof

and Matt a couple more friends, Charlotte's boyfriend Jens, and Brigit.

The dancing and drinking probably built up people's appetites just about the right amount, so by the time we said 'go' there was a near-stampede. The crackling was out of this world and we did lots of pork baps with apple sauce. A pig is a big thing and even though we continued to cook and serve for the next six (!) hours there was still some left. I can categorically say no-one went hungry (Brigit had done a big vat of vegan grub as well so we had all bases covered).

We did it!
At the height of everyone rushing for food, David Bailey tried to slip a £5 note quietly into my hand … When we had first seen the pigs and bought them from David he had confidently stated that there was no way we would eat them and they would end up living to a ripe old age at New House Farm. I must admit his confidence had been well founded, he knew that Brigit and Charlotte thought the pigs were great 'but not to eat'. It is a measure of how seriously the family is taking our lifestyle change that I was allowed to kill and eat them. Suffice to say I kept the fiver and then took it down to David in January to buy a sack of feed for the next bunch of weaners we got off him; he didn't lose it for long.

97

Keeping hens

You can't go self-sufficient without getting hens. They don't need much space. They turn scraps into lovely golden-yolked eggs. They can even provide more hens and possibly a rather tasty Sunday lunch. Anyone can keep hens and there is a selection of rather nifty pods available (check out www.eglu.com). If nothing else, the sums make sense. A hen can be fed for a couple of pence per day. A healthy hen can produce about six eggs a week. Free-range/organic eggs cost about £1.50 for half a dozen, so you can save over £1 a week – and you will know exactly where your food came from, with no food miles. You win all round.

Building the chicken run

We decided to get ourselves six hens. First we cleared a day in the diary and allowed everyone to get involved in building the henhouse and run. It was not intended to be a long or difficult day, but somehow our finely trained bunch of artisans took forever to make what was supposed to be a simple little henhouse. The inside run is the end of the potting shed with a rather clever little entrance door. The laying boxes can be accessed from the outside and there are sufficient feeders for all six hens to have one each. When they are in their cages, they have only three things to do: eat, lay eggs and peck each other. The idea of lots of feeders was to make sure there was no competition for food so they would have absolutely no reason to argue.

Our outside run was fully enclosed in 1-inch mesh to ensure that no foxes could get to the girls if we ever left them unattended for any length of time. I really feel that I must mention the door to the run at this time. Jonny the director of the TV series and Rich the soundman put a whole afternoon into making it and they looked rather pleased with themselves when they had finished. I would hazard a guess that it is probably the most filmed chicken-run door in history. I quite like the fact that it doesn't sit flush – it's quaint!

ESSENTIAL PREPARATIONS FOR GETTING CHICKENS

Must have	What? Why?	New House Farm checklist
Coop	Chickens need to be able to roost in a draught-free, airy, sheltered coop. It doesn't have to be large	Our first coop was an old packing case with a fireguard for a run. Palatial for six
	Somewhere to lay eggs (not where they roost or the eggs will always be covered in poop)	Our hens can access the boxes without going in the coop
	Must be fox-proof – they will kill any unsecured chickens	Fort Knox
	Door to outside run that can be secured	Door slides up and down a groove
	Door to allow entry/capture/egg collection/cleaning	Full walk-in coop
	Mesh of less than 1-inch preferable to keep the rats out	Yes, but rats have visited by day through the door
Run	Mobile runs with fresh grazing are great	10 × 15 feet
	If they are to be left unattended so they can't roam free range, the bigger the run, better	We went for a permanent large option
	1-inch mesh at least	1-inch mesh
	Buried 6 inches deep to stop the dreaded fox	Buried
	Door to allow entry/capture/cleaning	Full-size door
	Roof if the coop is less than 6 feet tall or if you ever intend to leave hens with open access from the coop	Our only weakness is that there is part with no roof (because of a tree growing through) so a very determined fox may be able to get in. We do secure coop every night!
Food	Layers' pellets/mash	We have available all the time
	Scraps, mainly carbohydrates (ours also like the caterpillars we find)	We share our scraps between the pigs and chickens
	Providing they are allowed to range freely and scratch around there should be no need to give them anything to keep the eggshells strong	Not an issue
	Access to areas that you have harvested vegetables from. They are great at scratching up the creepy-crawlies that can cause problems	Only if Anda is sure the vegetables still growing are safe for them
Water	Something that can't be knocked over	We hang the container from the roof
	Fresh every couple of days	

Our rescue hens

Brigit and I have kept hens on and off for years and James even had rare breed Speckled Sussex for a while, though they failed to make him his fortune. Brigit had heard that there was a charity specializing in rescuing battery hens and after a bit of digging around we had a long chat with Jane Howarth. She's a lovely lady who does great work resettling hens when they have finished their useful time in a battery farm.

I don't agree with battery farming but it is a fact of life at the moment, and Jane relies on the goodwill of farmers to be able to rescue thousands of hens.

It was a two-hour journey for us to get to Jane. We knew we had reached the right place when we encountered hens walking up the drive and saw several other cars with people clutching boxes and cat carriers. Then we spied hundreds of hens in large walk-in cages. It was lovely to see them sunning

How to rescue a battery hen

* Contact Battery Hen Welfare Trust: www.thehenshouse.co.uk.
* Find out when the next batch of hens are to be rescued.
* Build a henhouse and run.
* Get in a feeder and some layers' mash.
* Head off to Chumleigh in North Devon with a cardboard box with breather holes in it or a cat box and collect your hens.
* Pay 50p each for them. They will look somewhat bedraggled, but it is 50p well spent.
* Bring them home.
* Keep them away from other fowl for a couple of weeks until they prove they are fit and well.
* Enjoy lots of great eggs!

themselves and scratching around on solid ground for what was probably their very first time. We chose six that had varying degrees of baldness. As they were being caught and boxed for us, Jane showed us around. She is obviously a soft touch because all her hens have names and their own little story, and a large number have been nursed in her hen hospital.

We thoroughly enjoyed letting our six girls out into their palatial new home and we have relished their eggs ever since. If anything, the girls are somewhat over-friendly and like to come up to the house rather than go into the field. They are so used to being given titbits that they jump up and peck your shorts or skirt if they don't get something. Rocky (her clipped beak makes her look like a true pugilist) will even have a go at your toes. They had a tendency to head for either the back door or the lower vegetable patch, so we had to invest more fencing to restrict their movements. Now they have only about ¾ acre on our land, though I must confess to not blocking the holes in the fence that give them access to the additional 20 acres of field next door – we'd hate to have our land overgrazed by the six of them! Jane has got it right with her charity work. No one will ever convince me that our eggs are not a million times better for you than those from cage-kept hens.

Our hens have gone from having less than a square foot each in a cage, with no daylight and no solid floor, to a life governed by the seasons and in which they seldom reach a boundary.

DID YOU KNOW? As a nation we are getting better at buying free-range/organic eggs, but still 70 per cent of all eggs consumed in the UK are from battery hens. Most of these eggs are used in processed food and the catering industry. Check labels, or, in a restaurant or café, ask.

Nature in action

Not all our hens have survived. We lost a hen on Christmas day and surprisingly something similar happened on Christmas day about ten years ago. It must be that there are probably less people around and the predators get a bit cocky, or maybe it's the smell of turkey all over the place. The last time it happened, we saw the fox attempting to make off with 'Number 5'. (When we have chicks they get numbers until we know the sex – hens can then be named; cockerels are for eating). This time I couldn't account for one of the hens when I locked up on Christmas afternoon. We discovered the body at the edge of the

lower paddock. It was a bit messy, but the meat on the neck and head only had been eaten. It didn't look like the work of a fox to me, so I had a chat with some of the locals and the best guess we have come up with is a weasel, stoat or a feral ferret. It was very sad, but it is the price of being free-ranging.

Other poultry

We are being sensible and decided that we should learn to walk before we can run. I would have jumped straight in, but there are voices of restraint within the family. We have yet to explore fattening chickens for the table and raising a couple of turkeys or geese, but I intend to do it soon.

104

DID YOU KNOW? It doesn't matter what you do, a free-range hen will find somewhere obscure to lay her eggs.

Water Water Everywhere

Our stream is spring-fed in the village about 220 yards from the house. It has not dried up in living memory

Water potential at New House Farm

When we first saw New House Farm one of our initial impressions was that there was an awful lot of water around the place. The stream was in spate and looked like a little river, its path through the property was an unusual mixture of natural banks, concrete underground pipes and tunnels, and it disappeared at the top end of the lower paddock. But that wasn't all. There was a spring-fed well in the lower paddock that was completely independent of the stream, and Brian, the previous owner, had tested the water and found it to be fit for consumption (this involves two samples being sent away for analysis with a reasonable time gap between them). He thought the flow rate was about 200 gallons a day which, though not massive, was extremely exciting as it had the potential to meet what we thought were the water requirements needed to run our household systems.

A bit of history
A number of people have visited us to share information about the history of the house and our plot. We found that water featured a lot in their stories.

The priory of St Andrew, Tywardreath, was established in the village in the late 11th or early 12th century, but after Parliament enacted the statute dissolving the monasteries in 1536 it was razed to the ground and its exact location is a mystery. However, copies of maps and documents going back nearly two hundred years show that New House Farm was formerly known as Priory Farm, and there is evidence of orchards on the site.

At the northern end of the house there was a pond, probably a stock pond, where the middle section of the stream is now. (A pond would be a great addition to the

'wildlife corridor' and when we have time it is definitely on the agenda.)

One of our neighbours recently mentioned they had heard that the stream continued to contour to the end of the lower paddock and that there had once been a waterwheel there. Best guesses put the priory in the area of our upper paddock and the graveyard to our north. It could also be that the house is on or near the site of an old thatched pub called the Ferryman. We were surprised to find that, according to some old maps, the sea nearly reached our boundary in the 16th century. We'd never considered tidal power …

Water works

Lots of people go through life not ever having to worry about what is behind the light switch or what is inside a radio or a little box that does something clever. I, like lots of others, love to look in and play (and invariably break). Not content with changing our electricity to a green tariff, I wanted to go further and make electricity so that we could be completely independent. We had searched long and hard for a house that had a stream with a fall. We would use the stream to generate electricity to power at least some of our lighting. And we would harness the water from a well for our domestic needs.

About our stream

Our stream is spring-fed in the village about 200 yards from the house. It has not dried up in living memory. Jim and I saw the potential of the stream straight away and when we found out it was spring-fed, from a couple of springs in the village, we were ecstatic. We even had the added bonus of discovering that the stream swells after it has rained because the storm drains feed into it. We love the fact that the stream flows day in, day out, and can always be relied upon. However, it is not always well behaved. It has a limited flow until it rains, when all the local streets run into our storm drains and we have a serious amount of water. The culverted pipes are not matched and during a wet spell our water levels rise as the water backs up.

Unfortunately, our gorgeous, spring-fed stream didn't have the flow rate we needed to meet all our power demands, but, undaunted, we set off.

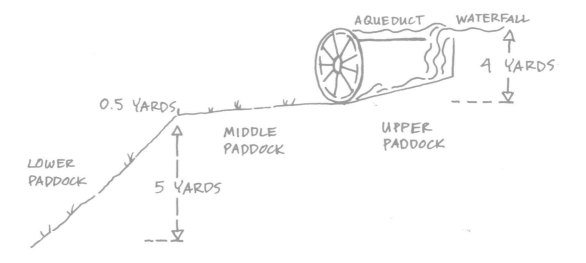

AQUEDUCT   WATERFALL

4 YARDS

0.5 YARDS

MIDDLE
PADDOCK

UPPER
PADDOCK

LOWER
PADDOCK

5 YARDS

The water route

The stream arrives at our boundary over a waterfall about 8 feet deep and it looks and sounds brilliant. It was always our intention to extract as much energy as we could from the water that passed through our property, so we mentally divided the drop into usable bits. The 'first drop' from the waterfall to our driveway was always going to be the easiest to exploit. We then had a relatively flat section that contoured under the driveway alongside our first vegetable patch and disappeared underground just as the land sloped away to form the lower paddock. This middle section was to be a wildlife corridor as, even though there was a reasonable flow, it was hard to find an easy way to harness the energy from the water and, as part of Brigit's permaculture, we also wanted to encourage wildlife.

Our configuration: stream in, waterfall, one 18-inch pipe underground opening into two 18-inch pipes under the road, an open stream, an 18-inch pipe under a track,

an open stream and, finally, the real problem area, a 10-inch pipe under the lower paddock.

Wheel or turbine?

We had a couple of options for making power. We could feed the water through a pipe and drive a turbine, or we could make a waterwheel. The turbine was a more efficient solution, but it did not have the ability to harness the extra potential of the stream in spate, and it would be rather ugly. So we decided to build our very own heritage-wheel-of-the-future. Never mind a piddly little pond and a fountain feature, New House Farm was going to get a 4-yard wheel and a 40-yard aqueduct.

Building the wheel itself was all about planning. We knew an all-metal beast would be too heavy, so we decided to go for a metal hub and make all the rest out of wood. The planning phase involved drawing templates for each of the component parts and making them all before the grand

Lots of people go through life not ever having to worry about what is behind the light switch. I, like lots of others, love to look and play

assembly. We visited a local engineering firm and bought a bit of pipe that went over our silver steel shaft and welded plates on to it at each end so we produced something that looked a lot like a very large, very thin bobbin. When the plates were marked out, we attached the spokes by drilling holes through the plates and bolting them on. The buckets were all made out of cut sections of plywood that we screwed on to the spokes.

112 If there is one key piece of information that I would give any would-be waterwheel builder, it's that the buckets need to be at 114 degrees. I'm sure there is some very complex theoretical derivation for the angle of the bucket, but Jim and I found the number so often, from lots of different sources during our research, and it looks right on ours! Armed with that little gem of advice, I'm sure anyone could build a waterwheel.

Building the aqueduct

Having just about satisfied Brigit on the domestic front – things were growing, we had provided the basics for life as we know it, it was September before Jim and I were allowed to take the team to the upper paddock and play. Everyone mucked in at once and after a significant amount of measuring and surveying we had decided on the route of our aqueduct and where the posts would go. Day 1 was surveying, measuring, cutting, digging postholes 8

and 12 feet apart to allow us to use the sheets of plywood we had, and finally cementing in our uprights. Personally, I think we put an awful lot of faith into the £15 laser spirit level we'd picked up at a car boot sale.

Day 2 and the bench saw was taken out of the workshop and screwed to the hard standing. We then proceeded to build our framework and our trough. As the light was fading we finished sealing most of the joints. We realized we needed to let the water flow to see if it was actually uphill, as it appeared to be to the human eye. As we were putting on the finishing touches, someone managed to magic up a tiny model of a dog and a surfboard to ride along the soon-to-be-released torrent of water. No one thought it at all unusual to have a surfing dog test our aqueduct ...

Everyone was asked to stand at the

point on the trough they thought would be the limit of the water's travel. Jim and I were the only two of the ten present who got it right, because we said it would make it to the very end, and indeed it did, despite leaking most of the way down – o ye of little faith! The leaks were mere teething problems, soon sorted with a tin of emergency repair tar.

Passing water

You have a number of different ways of passing water over your waterwheel. Overshot is the traditional solution. As we had made our wheel as big as possible, there was very little clearance between the bottom of the wheel and the water level. If the stream is in spate, the water might come up to the bottom of the wheel while the extra water will pass under the wheel and then have to come back to join the

stream. There will be a point under the wheel where the water is trying to go in two directions at once. The resulting turbulence will undoubtedly cause us to lose power.

Having discounted breastshot (where the water strikes the buckets on a huge wheel about halfway up, causing the wheel to rotate and then the water flows out underneath it), our chosen solution was a backshot arrangement. Our nearly horizontal aqueduct minimizes the forward momentum in the water. The total amount of water that arrives at the wheel is limited by the size of the trough. When the water reaches the end of our aqueduct it falls through a trapdoor and causes the wheel to rotate anticlockwise. The only thing that is a bit of a pity is that when you come along our drive it appears that the wheel is going backwards!

I love the sound it makes,
the fact that it turns day
and night and the way
it is harnessing the spring
water and the rain –
everybody should have one!

Gearing system

If we are not trying to make electricity, our wheel turns at 20 revolutions per minute (rpm). In other words, that's our no-load speed. However, when we connect the wheel to a generator to make electricity, it is slowed down to about 6rpm. Our very efficient generation device (a permanent magnet alternator, or PMA) was designed for use with a wind turbine and works best at higher speeds. We therefore needed a gearing system to increase the wheel speed of 6rpm up to the optimum for the generator of about 300rpm – about 50:1. I'd rather be lucky than smart, and a conversation with Mike, who owns the smallholding behind us, uncovered the fact that he had a gearbox in his back garden that he'd put under a sheet of polythene 20 years earlier.

It was a 44.2:1, oiled filled box, which, once the old 110V electric motor had been cut off, took only an hour to free up. OK, 265rpm is just under what we were aiming at, but beggars can't be choosers.

We had the means of making electricity. It was time to connect it all together. Within a day, Jim and I had a rather professional-looking power plant inside the lean-to beside the wheel. Even though we had a teething problem – in that we had to nudge our wheel back into place and fiddle with our couplings – that evening we thoroughly enjoyed drinking a beer by the light of a bulb that we were powering from our wheel.

The waterwheel makes me smile every time I come into the driveway. It is the first thing people see and it definitely makes an impression.

HOW WE CALCULATED OUR ELECTRICITY OUTPUT

Flow rate (Q) average/low (high rate not yet known)	2 litres per second = 0.002 cubic metres per second
Drop available at the end of the aqueduct, the head (H)	3.6 metres
Efficiency of the generator and wheel (k)	0.6
Power available = Q × H × 9810 × k	43W

*Note for the techies: 9810 = density of water multiplied by the acceleration due to gravity. 43W is the minimum expected but it should be noted that it will be generated 24 hours a day and stored in batteries so it can be used on demand.

Some of this may seem complicated, but it is surprisingly easy to get help with bits of engineering

Taming the flow

The supply of water is reliable in that it is always there. All it takes is a little rain and we have a lot of variation in the quantity flowing down the stream. We have had some cracking downpours that have produced torrents that filled the system to capacity. Jim and I couldn't believe our luck, and whenever it rained hard we could be found walking around in our wellies smiling at the ferocity of the water. Lots of people like walking in the rain (don't they?) but it is much better if you can appreciate that extra rain means extra power.

The size of the aqueduct limits the amount of water that can be delivered to our waterwheel. We made the aqueduct a lot bigger than was necessary to handle the flow we had seen in the summer months as we hoped to get more water and, therefore, more energy, in the winter when we would need the lights on more often. We matched the aqueduct and the buckets in the wheel for width and left the speed up to an educated guess. We didn't do badly, and even when the aqueduct is full the wheel speed is not excessive when under load.

Some of our little problems

First problem was losing a couple of teeth in the gearbox and sheering the connections, so we attached the 'half a gearbox that was left' straight on to the alternator (not optimum revs, but we did make electricity). Next catastrophe: we had a downpour, and one of the couplings sheared (an 8-millimetre shear pin snapped). Fortunately, we managed to track down a scrapped variable gearbox which we then attached to our half gearbox and we had an adjustable, optimum electricity-generation system. It was obvious that the waterwheel was powerful enough to snap bolts, shear pins and gear teeth, so we went back and strengthened all our couplings. Touch wood, we are now fully operational.

When the waterwheel is under load (working hard making electricity) it turns at about 6rpm, which is sedate, but it gives the impression of lots of power, making its rhythmic whooshing noise as the buckets fill and then empty.

When the alternator is not connected or the gear train fails the noise and the feel of the beast are very different. It rotates at over 20rpm and hums in quite a menacing way. As soon as this happens it is far from relaxing to listen to it, and it is best to turn it off and investigate the problem. I would go as far as to say it gets a bit scary and looks as though the wheel could really hurt if anyone was silly enough to touch it (which they can't because there is a fence to stop them). However, it is not damaging itself, which means there isn't even a real urgency to turn it off other than the fact that we have to pay for our electricity when it's not working.

THE TECHNICAL BIT

We rectified our three-phase output from the PMA. (Build your rectifier yourself – Jim bought the bits for about £6 and it took half an hour to put on the sheet of aluminium.) The DC output was paralleled to our charge controller, the batteries (four 12V 110 Ah deep-cycle leisure batteries connected in series and parallel to provide 24V) and our pure sine wave inverter. As we intended to run our lights, a pure sine wave was essential, even though it is more expensive. There was a faint chance we might generate more power than we needed, so we had to have a dump load on our charge controller to stop the batteries from frying. For this, we used the heating element of an electric fire.

When you make your own electricity,
you become religious
about turning lights off

Results

Some of this may seem complicated, but please don't shy away from what can be a lot fun because you think it is too difficult or because you have never done it before. It is surprisingly easy to get help with little bits of engineering. We don't yet have a workshop with lathes and large metal facilities, but there are lots of firms out there who do. For example, Hydra SW Enginering – Indian Queens (so named because Pocahontas stayed there – how's that for some trivia!) were great when we couldn't get our shaft to fit in the bearings we had 'procured'. They sorted it in minutes and were always happy to give advice.

It was obvious, after having evaluated our water flow, that we couldn't meet all our needs with a waterwheel. So we decided we'd use it to power the part of our house that could be easily isolated from the grid and which used approximately 1.032 kilowatt hours (kWh) per day (43W × 24 hours): our lighting rings. This overcame technical issues about connecting the waterwheel generator to the mains (currently there is no DC inverter approved for connection to the grid).

We did the sums using the power needed for low-energy bulbs over a 24-hour work cycle as dictated by the Strawbridge family. The sums are approximations; not all the bedrooms have the lights on for an hour, though some may have them on for more, and two bathroom hours a day is more than enough sometimes but it does depend on who is staying. When you make your own electricity, you become religious about turning lights out.

It would appear that, with our 1.032kWh per day, we were just a little bit short of juice... I never did tell Brigit but I think I got away with it because when we need more lighting – in winter – there is more water available and consequently we get more than our miserly 43W. And, as the days lengthen in summer, the amount of time you need the lights on decreases. It does pay to have a fallback plan. Where the power cable comes on to the lighting distribution box, we have a throw switch which, should we run out of power, can either reduce the amount of lights on or go back to the grid.

LIGHTING NEEDS (using low-energy bulbs)

Room	11W	20W	Total	Hours	kWh
Bedrooms	7	0	77	1	0.077
Bathrooms	2	0	22	2	0.044
Lounge	0	2	40	4	0.16
Hall	0	2	40	5	0.2
Landing	3	0	33	1	0.033
Kitchen	0	4	80	7	0.56
Utility	0	1	20	1	0.02
				Total	1.094

Success

All in all, we are very happy that we have harnessed the power in the upper section of the stream and it looks great. We set aside a day to clear the watercress out of the middle section, which allowed the water to flow freely. It did look a little as though we had implemented a scorched-earth policy, and Brigit had a hairy fit because we had violated a wildlife corridor (we are fortunate that the slug and snail population is predated by lots of frogs and toads that lie along the stream, and we even had some wild ducks when we first moved in but I think a couple of days with Tex, Anda's terrier, was too much for them and they were never seen again). When Brigit calmed down I managed to convince her that we had left a complete bank untouched, and that the water flow was essential to ensure the stream didn't silt up as it would increase the likelihood of the stream breaking its banks and flooding the area of the pole barn and the back of the house. In addition to clearing the stream, we decided it was necessary to act before we found out exactly how bad the problem could be, so we dug out a little channel to ensure that if the water level rose too high it could escape down the valley out of harm's way. You get a smug feeling when you see a cunning plan coming together – the overflow channel has saved our bacon a couple of times, which makes all the work worthwhile.

Total costs for the waterwheel system

I'm always interested in the bottom line, so I've included our costs and where we sourced our components. Labour is not included in the costs – we had so much fun I considered charging people for coming out to play!

The cruel facts are that, if only minimum flow was available and electricity prices stayed the same, it would take over 25 years for the wheel to pay for itself. But somehow I think electricity prices are going up, and I have yet to tame all the extra power I get throughout the winter, so I'm sure that payback will be a lot sooner. At its peak, the system produced a rather impressive 386W and, over the winter months, never produced less than about 80W (twice our forecasted power). We even left the Christmas fairy lights on the old, dead cherry tree in the upper paddock and powered them with our spare (!) energy.

Building and owning a waterwheel is about a lot more than getting cheap, sustainable electricity. I love the sound it makes, the fact that it turns day and night and the way it is harnessing the spring water and the rain – everyone should have one!

COSTS FOR THE WATERWHEEL SYSTEM

Item	Cost (£)	Source
Aqueduct		Mainly local Travis Perkins
Posts – tannalized sawn 4 foot × 2 foot	200	
Trough – WBP 9mm ply	150	
Cement	30	
Waterproofing	40	
Wheel		Mainly local Travis Perkins
Hub	120	Local engineering firm for metal
Spokes	50	
Buckets – WBP 9mm ply	90	
Waterproofing	20	
Bearings	75	PSL @ Bodmin – really helpful
Silver steel shaft	Purloined!	Skimmed to fit by the blokes at Hydra SW Engineering – Indian Queens
Gearbox	Free	Supplied by Mike and Angie next door
Generator		
Universal joint	Free	Peter Trebilcock helped make this
Permanent magnet alternator	150	EFX, best deal we could find on eBay
Batteries	160	Par Batteries and Brakes
Pure sine wave inverter	80	Wind and Sun (www.windandsun.co.uk)
Charge controller	40	Wind and Sun
Cables	40	Dave Smith Electrical – ace bloke
Connection to the lighting circuit	60	Dave Smith Electrical
Throw switch		
Grand total	**£1,305**	

Every year in the Malverns the summer solstice is celebrated by 'dressing the wells' when each of the many wells around the hills is decorated. The ceremonies are worth a visit and they remind everyone that the springs used to be essential for all those living there. We'd been in Cornwall for less than a week and there were a dozen things crying out to be done, but we now had our own well and couldn't resist starting a new dressing tradition at New House Farm. The decoration was simple and very natural, but it was worth the celebration. Above all it reminded us that the well was there to be used, although it took several months before we had the opportunity to spend the time necessary to exploit it.

Domestic water supply

Water for the plants

Our domestic water supply is metered and part of the reason our first water bill was so high (£178 for the first six weeks) was that we had to water the newly planted vegetables until they were established. We didn't really think about the cost, we just used the hose because we needed to. Obviously, irrigation using the stream had to be the way forward. Pouring expensive, purified, chlorinated tap water on to the plants made no sense and changes had to be made. The first, short-term solution was to push one end of a hose pipe into the stream at its highest point and take the other end to the beds. It was very simple and it worked. There were minor issues of dead vegetation creeping into the end of the hose to block it and a lack of pressure compared to the mains, but the water was free, which made the inconvenience bearable for a while.

However, we decided that watering took too long, so we built a platform from scraps of wood and ply about 6 feet high and put a water butt on it. To fill this we used water from the stream, which we collected in a dustbin and then pumped up to the butt using a very low-energy pond pump. It cost pennies to run and increased the pressure, but it did involve running a mains electricity cable to the top paddock, which wasn't frightfully elegant and could only ever be a temporary solution. What we needed was a reliable, pressurized, free water system, so that was what we built.

Water use in the house

Funnily enough, the catalyst that launched us into using our well was the bill for the next three months' water: £450! We were gutted. Then we took stock of the amount we had been using and it all made sense. The house regularly had ten or twelve people living in it with an extra four or five working there. We made the decision from Day 1 that anyone working with us was invited to lunch, or if they had travelled a long way they were given the opportunity to stay with us. This was not purely altruistic: we learnt a lot from them.

125

With so many visitors staying with us and using our facilities, Brigit was forever changing the bedding. A couple of years ago, we even invested in a huge Dyson double-drum, monster machine to allow more washing to be done at once. When we did the research as part of our desire to go green I was relieved to discover that it was efficient as well as large, as I had a terrible fear that we would have to get rid of it and buy a tiddly little eco-machine – it would have been murder for Brigit!

It is not just the washing machine. Lots of people use lots of water: there are showers, the very occasional bath, cooking, washing all the pans and, probably a huge contributor, the loos. Dishwashers are greener and save energy compared to washing dishes by hand (in clean warm water and rinsing them) – hooray! Using the dishwasher is great, but it should be completely full (which for us was after only one meal) and you should use the 'eco-wash' programme but it's great to know that being an eco-warrior doesn't mean your hands have to get wet and wrinkly.

Using the well

We were fortunate enough to have the children and their friends pop home from university for a long weekend in the autumn. An addition to the already long list of jobs was digging a small utility trench for pipes that would allow us to pump water from the well to the house. While we were digging it, Jim was building a nice little shelf and door over the well entrance to act as our pump house. The trench was to be only 18 inches deep, but it wasn't a small one. The first 44 yards was through the area where we had initially kept the pigs. This had become very boggy and thoroughly unpleasant. I'm sure the chaps loved digging there as much as I did and it only took us about an hour's hard graft. That was the easy bit. The final 27 yards to the potting shed was through compacted, rocky ground; it's fair to say, senses of humour wore a bit thin, but we did get it done.

From the potting shed it was all plain sailing. We had found a route from there to the attic that provided a reasonable amount of frost protection, not that it is ever that cold in Cornwall, and we set about laying the pipes. From the well to the potting shed was 1-inch outdoor piping and from the potting shed to the attic was half-inch piping (we didn't need more than half and inch as we intended to fill a large tank in the attic to ensure we had a reservoir and the ability to use lots of water at once).

Pumps, pipes and powering the tank

In Jim's previous life aboard his canal boat he had used a small pump that produced water on demand. Our biggest problem was to have sufficient head to reach a tank in the attic. Water is heavy – 1 kilogram per

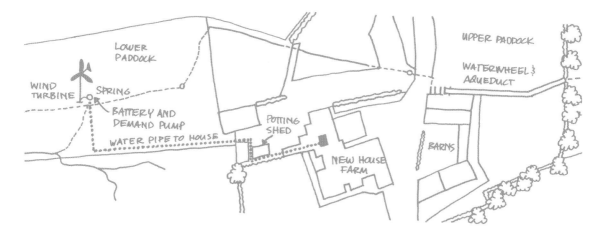

litre to be exact. To pump it to a higher level than a pump, the pump has to be powerful enough to push a weight of water above itself. The height it can achieve is the 'head'. We were asking a pump to push the water from halfway down the lower paddock up into our attic to the top of a tank, and to deliver enough to keep our tank full; a fairly significant feat. In comparison, the pump for a water feature usually has a head of only a couple of feet. After the obligatory research we ended up with a battery-powered 12V Flojet pump. It had the advantage of being able to run dry if we ever emptied our well faster than the spring could fill it, and was capable of hundreds of gallons an hour. The first test we did involved attaching a hose-pipe to a bamboo pole on the top of our scaffolding tower, so that it was higher than the house, and laying it all the way down to the well. We weren't sure what to expect but, fortunately, it was only a couple of moments before we had a geyser. Sometimes if you are at the limit of a pump's head the throughput may be only a very little flow. We couldn't believe our luck. As if this weren't good enough, we also discovered that the pump, though working furiously, was not even making a dent in the amount of water being supplied by the well, which was producing hundreds of gallons an hour, not a day – I know I've said it before, but I'd rather be lucky than smart.

After this success we were satisfied that we had all the components we needed; it was just a matter of putting them all together and begin saving masses on our water bills.

All you need to get free water

⬥ A source of water – the cleaner, the better.
⬥ A trench connecting the source to your in-house storage tank.
⬥ Piping.
⬥ A pump.
⬥ Power.
⬥ A tank in the attic.
⬥ A cunning plan.

The changeover

Before the changeover all the water in the house was supplied by the mains. There was a header tank for the hot water but everything else was directly fed. We wanted to reduce our reliance on the mains to the minimum yet keep its supply as a backup. The kitchen sink and the sink in the utility room would be mains-fed because it was easiest to keep them linked. All the loos, the dishwasher, the washing machine and showers would be fed from the well. In addition we wanted to use the well to suply all hot water. A requirement of the changeover was that we would be able to feed mains water into the tank if necessary.

After a walk around the plumbing we found the best places to disconnect from the mains and swing over to the well-fed system. Our tank was placed as high as possible in the roof space, which involved building a little platform for it between the main A-frames. We then connected the tank to the well output and put in place all the connections we thought we could possibly need. Preparations in the house took the best part of a day. However, once we had double-checked the tank connections we didn't need any encouragement to switch to the water from the well – it was a matter of connecting the battery to the pump in the well and a couple of minutes later water was gushing into the tank. It all went

smoothly except that, even with the tank raised that extra couple of feet, there wasn't quite enough force to operate the thermostatic valves in the shower properly. This was soon rectified by putting an in-line pump on to the hot- and cold-water supplies. Now that the system is in, you'd never know we don't have conventional plumbing except, of course, for the reduced bills.

Poop and waste sewage

About two-thirds of the cost of water is sewage charges. Cutting down on the amount of processed water you are using makes no sense if you are still filling up the drains and sewage system. When we arrived at New House Farm we had every intention of coming off the mains sewer and building our own reed-bed system. We like the balance this kind of system achieves, and it would be great to know that our smelly waste was not a problem we were passing on to the water company. A trip to the Centre for Alternative Technology (CAT) changed our minds. Although I was impressed by the fact that the systems they had built were smaller than I had expected, discussions with the experts led us to question whether the amount of effort it would take to set up our own system would be balanced by the gain we would achieve. We were convinced that the cost gain was not enough to make it worth doing.

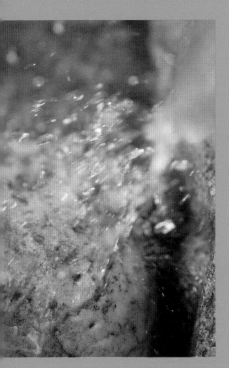

BOTTLED WATER

When we lived in Malvern we used to drink water from the hills and a local initiative encouraged people to put a small contribution into a charity box for Wateraid every time they collected some. I liked the logic and ever since I have been keen to support charities that promote the provision of water in developing countries as I believe that a clean water supply can add to the health and dignity of a community. Clean water is something we take for granted. That said, we still have an innate, very British, concern that when we go abroad we will end up with the screaming abdabs unless we drink bottled water. In some parts of the world this is very true – when I visited James during his gap year I had a particularly bad time in Kathmandu after some ice that I was assured had been made from bottled water obviously hadn't been. Why not take a look at FRANK (www.frankwater.com), which offers non-profit, natural bottled spring water. It aims to supply the UK's demand for bottled water in a socially responsible way – by giving the net profits (up to 60p in the pound) to the clean-water projects that it funds. If you are going to buy bottled water anyway, why not drink FRANK and help to bring about positive change in some of the world's under-served communities, as well as making yourself more appreciative of what comes through your taps.

DID YOU KNOW? Less that one per cent of water treated by the public water systems is used for drinking and cooking.

A composting loo

The experts had steered us away from a reed bed, but we still had to try to reduce our sewage. It was time for a compost loo. Anyone who was in the army and has been on exercise in Germany has probably experienced the 'long drops' on Sennelager training area. Twenty-five years ago it was a large building with a set of loos along the outside wall. It was very Romanesque in that there were no cubicles and we all sat together. The most unusual aspect of the loos was the complete lack of water/flushing. They were literally 'long drops'. I have absolutely no idea why they never filled up after years of use (there must have been a gradual seepage away), but hundreds of soldiers used them after days on 'compo' rations. It just doesn't bear thinking about! Composting loos are very similar. Sometimes called biological toilets, dry toilets or waterless toilets, they contain, and control the composting of, poo and toilet paper. Unlike a septic tank, they rely on un-saturated conditions – material is not immersed in water. If operated properly, a composting loo doesn't smell and breaks waste down to 10 to 30 per cent of its original volume. The resulting end product is a stable, soil-like material called 'humus', which can be used as a soil conditioner.

The main task of a composting toilet system is to contain, immobilize or destroy organisms that cause human disease (pathogens), without contaminating the environment and, more seriously, without harming the locals – in our case, us! This has to be accomplished in a manner that is consistent with good sanitation and doesn't attract flies. It must also produce an inoffensive and reasonably dry end product that can be handled with minimum risk and minimum odour. The ideal composting toilet transforms waste into something that can be used on plants.

Build your own composting loo

Siting has to be convenient yet unobtrusive. We chose behind the cider shed, which is well placed for the gardens, upper paddock and outbuildings. For 'batch processing' the composter/reactor needs to consist of two discrete containers to allow you to alternate between them. Some form of drainage is necessary in case liquids build up. You also need to separate urine from solid waste – they are treated differently. It's a fact of life that when you sit down, boys and girls pee forward. A carefully placed, curved bit of galvanized metal with a 'gutter' at the bottom catches the majority of the liquid and keeps the 'compost' dry. In our case, Anda wanted the urine for our compost heaps. Whatever you do with it, there must be an access door for removal of the end product.

Ventilation must be flyproof and provide enough light to encourage any flies in the

On the sign in the image:

PLEAS[E]
ADD A COUPL[E]
[O]F SCOO[PS]

How to use your composting loo

* It's not for 'number ones' but when you use it for 'number twos' make sure any pee goes in the direction your design intends it to.
* Do not use too much paper.
* Pour in a mug of sawdust after use.
* Always close the lid after you have finished to keep flies out.

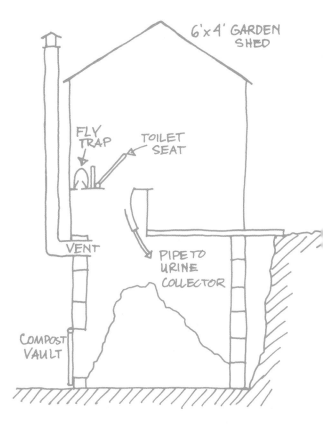

reactor to head up to the dead end and die. It also allows the oxygen needed by the aerobic organisms in the reactor to enter. It is important to fit the loo seat tightly to prevent flies entering or leaving the reactor.

Compost loo convincing

Before we moved to Cornwall James and I were chilled about the idea of a compost loo, but Brigit and Charlotte were yet to be convinced. By the time Brigit joined us after her permaculture course she had used one and had been pleasantly surprised. With only Charlotte to come round to the idea, Jim and I set out to build a composting loo. Neither of us is a brickie but we built the supporting wall in a sturdy enough fashion to make sure it would survive even the heaviest of visitors. Failure to do so would have meant someone literally ending up in the sh*t! Making the shed floor was again a matter of overkill. We mounted our vent and built our shed and decking area without issues. The internal functioning was easy and I reckon it was quite palatial. Looking

at the finished product, we were very aware that it had a huge capacity, and that it would take a serious amount of lentil action to fill it. One thing is for sure. If we prevent 6 litres of waste water entering the sewerage system each time it is used (the long flush on the eco-toilets), we have created something that can reduce our waste for years to come. More to the point what did Charlotte think? …'It's not as bad as I thought it would be, but why would anyone ever choose to go to the loo outside?'

The ideal composting toilet transforms waste into something that can be used as a soil conditioner for plants and trees

Rain-water harvesting

Over the last hundred years people have grown to expect a reliable, clean and inexpensive water supply, but before that rain water was routinely captured, so harvesting it is not a new idea. The size of the storage tank is determined by the amount of water available for storage (which in turn depends upon the roof size and local average rainfall), and the amount likely to be used (which depends on how many people live in a building and what the water is used for).

The great thing is that every time you use stored water, you save mains water and slow down the ever-spinning meter. In addition, you will have stopped storm water from entering the drainage system. Such savings are becoming more significant. In recent years, dry periods followed by heavy rain have frequently led to serious flooding. In our village there has been a large investment in managing storm water and the main road was dug up for weeks. In some instances stored rain water can provide an off-mains supply in remote areas and addressing

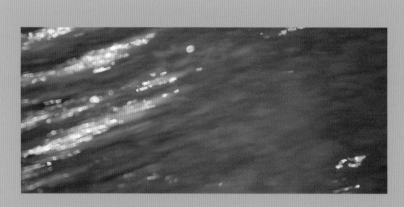

ULTRAVIOLET STERILIZATION SYSTEM

I was surprised to discover that the Water Supply Regulations, 1999, classify rain water in the same category as raw sewage. This may at first seem unreasonable, but the rationale is that it may be contaminated by bird or rodent droppings from the roof and so may contain harmful organisms. The bottom line is that for rain water to be considered fit for drinking it must be treated. One of the most environmentally friendly ways to do this is through ultraviolet sterilization. The rain water you intend to drink must be filtered to remove any sediment before it enters the unit, as bugs can hide behind little lumps of sediment and thereby save themselves from being neutralized. Once fitted, in line with the pipe supplying the drinking-water tap, the system is safe, efficient and simple to maintain: the ultraviolet lamp has to be replaced annually, and the cartridge-type sediment filters need to be changed periodically depending on how dirty the water is. The systems are not cheap, and I know someone who has been using his own home-made one for years and is none the worse for it. If you intend to use water that is in any way risky an ultraviolet sterilization system has to be a great investment.

When you try to live a low-impact lifestyle it is fundamental to ensure you manage your water effectively. Water shouldn't be wasted and, whenever possible, it shouldn't be over-processed rain-water harvesting is increasingly becoming a requirement of the planning system.

When you need the water, all you have to do is pump it to where you want it. It is worth making sure you have the correct filtering in place, that there is an effective overflow arrangement and that the plumbing from your harvested tank is kept separate from the mains.

Water storage

A rain-water harvesting system need not be expensive. It's just a matter of collecting the rain that falls on to roofs, then storing it in a suitable tank until it is required.

Let's face facts. It rains a lot in south-west England. Cornwall averages about 1 metre of rainfall a year. When you combine this with the fact that we have 230 square metres of roof, there is a significant amount of water that we could harvest. And with our local water authority charging over £3 per cubic metre for water, we couldn't ignore what could easily be collected for free – especially after we had gone to all the trouble of investing in a nice new roof

135

A basic rain-water harvesting system

◊ Normal guttering and downpipes, to direct the water.

◊ A storage tank (frequently underground) to hold the harvested water, which is usually filtered on entry. Check out the Wisy filter which ensures that about 85 per cent of the water is diverted into the tank while allowing the debris and remaining water to drain away in the normal manner. When I first saw one at the Centre for Alternative Technology I couldn't figure it out, but it does work.

◊ A highly efficient and reliable submersible pump to deliver the water on demand or deliver it to a normal header tank. This is a bit like the system we used for our well water.

and guttering. When we get around to reno-
vating our outbuildings it will be extremely
useful to have a supply of water available
right there, so we will be looking at harvesting as much as we can. We decided that the
most useful way to use the rain water was to
collect it in water butts at a number of downpipes where it can be immediately tapped
into watering cans and used for plants, or the
animals, or cleaning, or mixing cement. We
have found lots of uses for it, but the best
part is that water is available all around the
plot so we never have to carry it very far.

We know exactly how lucky we are to have
a well full of spring water and a stream that
never runs dry – and we fully intend to use
these assets to the best of our ability.

Our way forward

It was after we had committed to, and built,
our wheel that we met Peter Trebilcock of
Valley Hydro. He lives in the next valley and
has forgotten more about water turbines
than I'll ever know. He's a real engineer: if
he needs something he'll make it; he has a
workshop to die for; his house and workshop
have never been on the grid; he uses all types
of sustainable energy – and to top it all his
wife Val is lovely too! I still believe our wheel
was the right answer for New House Farm,
but I'm now keen to build and use a water
turbine as well. I'm trying to see if I can
justify building one for the lower paddock.
Jim and I picked up a cheap little water
turbine to play with. It is not as picturesque
as a waterwheel, but it is very compact and
all we have to do is sort out the piping to

Fuel and Travelling

There is an old German saying I like: 'Only a dead fish goes with the flow'

Low-impact lifestyle

When trying to live a low-impact lifestyle, you are faced with decisions every day. We use fuel for almost every activity we undertake: cooking, driving, working (lights, heating), even leisure. Those with healthy pastimes use fuel-energy too: ramblers, hill-walkers and golfers tend to drive to the start point; gardeners love visiting garden centres ... In short, it is fair to say that living in Britain uses fuel. The only place I have been to where I felt it was possible to live without fuel was Trinidad, but that meant subsistence-level living, and the climate there is mild.

In some areas we have to make explicit decisions about what sort of fuel to use, such as for personal transport and heating. In other areas our decisions have an implied impact. Our electricity provider uses fossil fuel or nuclear power unless we specify a green tariff (if you haven't changed provider yet, do it now). We have the opportunity to be proactive and make positive decisions rather than just going along with the majority. There is an old German saying I like: 'Only a dead fish goes with the flow.'

Fuel for thought

🔥 There is a finite amount of fossil fuel on our planet. I don't want to debate how long it will last, but there is definitely a limited supply.

🔥 Emissions from using fossil fuels are having a detrimental affect on the atmosphere. Some people are in denial about global warming, but it's clear our outer atmosphere is changing for the worse.

🔥 Public transport seldom takes you from the start of your journey to your destination.

A little lesson on fuel

To understand biodiesel, it helps to know about the volatility of fossil fuels. The most volatile is gas, followed by aviation fuel (avgas), followed by petrol, then diesel; the least volatile is engine oil. It is well known that diesel doesn't ignite easily compared to petrol and avgas. What is not so well known is that it is possible to reduce the volatility of avgas so that it can be used in lieu of petrol by adding oil to 'dampen' it (avgas being much cheaper than petrol or diesel). It is illegal, dangerous and foolhardy, but it just goes to show that there are ways and means of blending to achieve a usable fuel.

Any fuel made from vegetable oils or alcohol needs to have the same characteristics as the fossil fuel it is replacing. The trick is to process it properly. I really like the idea of growing a crop, harvesting the oil and producing biodiesel from it. I suppose it all comes back to nature doing its bit and us benefiting. The description 'liquid solar energy' is particularly fitting and appealing.

What is biodiesel?

Biodiesel is an alternative fuel with similar properties to conventional diesel. It has other uses – such as in boilers – but we considered it only for running our cars. Beware of the hype: when you hear a motor fuel is 'biodiesel' it might be 100 per cent pure or it might be combined with

conventional diesel in a proportion as low as 2 per cent. Always check!

Biodiesel is composed of free fatty acid methylesters (FAME). It can be produced from vegetable oil, animal oil/fats, tallow and waste cooking oil. Virgin vegetable oils could also be used, such as rapeseed, palm or soyabean. In the UK, rapeseed represents the greatest potential for biodiesel production, but it is not being produced commercially because the raw oil is too expensive. After the cost of converting it to biodiesel has been taken into account, it simply cannot compete with fossil diesel – it's just too costly. It is for that reason that most biodiesel is at present produced from waste vegetable oil sourced from restaurants, chip shops, industrial food producers and so on. I was fortunate enough to find a chip shop in Mevagissy, a local seaside town, which had waste oil to be disposed of. Naturally, I volunteered to take it off their hands.

Biodiesel is a relatively clean fuel and does not contribute to the causes of global warming. It is possible to find sources of biodiesel in the UK, but the industry is still in its infancy.

141

What are the benefits of biodiesel?

● Biodiesel is rapidly biodegradable and completely non-toxic, meaning spillage represents far less of a risk than that of fossil diesel.

● Biodiesel has a higher flash point than fossil diesel and so is safer in the event of a crash.

● Biodiesel reduces tailpipe particulate matter, hydrocarbon and carbon mon-oxide emissions from most modern four-stroke compression-ignition engines. The reason is mainly the presence of oxygen in the fuel, which allows more complete combustion. The higher the proportion of biodiesel in a mixture, the more the noxious exhausts are reduced.

● Biodiesel has a very low concentration of sulphur compared to traditional diesel oils, and about the same as the amount in ultra-low sulphur diesel (ULSD). (Sulphur can lead to the pro-duction of sulphur dioxide and, as a result, acid rain.)

● Biodiesel improves the lubricating power of the fuel and therefore ensures correct operation of key components in the engine system.

How we made our own biodiesel

First I have to confess I haven't done chem-istry since my 'O' levels, but here goes …

There are three routes to biodiesel pro-duction from oils and fats:

● Base-catalyzed transesterification of the oil.
● Direct acid-catalyzed transesterification of the oil.
● Conversion of the oil to its fatty acids and then to biodiesel.

However, almost all biodiesel is produced using base-catalyzed transesterification because it is the most economical process, requiring only low temperatures and pressures and producing a 98 per cent con-version yield – more than enough for us.

It is possible to use ordinary vegetable oil as a substitute for diesel, but there are problems such as relatively high viscosity leading to poor atomization of the fuel, incomplete combustion, ring carbonization, and accumulation of fuel in the lubricating oil, unless you modify your engine with a fuel pre-heater. If you do pay for a pre-heater you also need to know that in 2006 the government ann-ounced that tax of approximately 47p per litre would be payable on vegetable oil used in vehicles.

The process

This is probably one for the 'don't try this at home' category as it needs expert knowledge or as we did, some serious research.

FILTERING

The first step was to filter the used chip fat to get rid of the lumps of fish, etc. (through a sieve, then through a little filter bought off the internet). We next dried it to reduce the water content by passing it through a sealed container of silica beans.

MIXING OF ALCOHOL AND CATALYST

A small test was done on the raw oil to calculate the required amount of catalyst. This was then dissolved into the appropriate amount of methanol. We had to be careful here as this was probably the most dangerous part of the process – sodium hydroxide and methanol are highly poisonous and not very nice. I got Jim to do this part.

THE REACTION

The oil was put into our home-made reaction vessel (an upside-down, old domestic hot water tank) and heated to about 50°C. We used an old central heating pump to keep the liquid moving, then we slowly added the alcohol/catalyst. (Our cunning system sucked it in through a side pipe and we didn't have to touch it!) The whole lot mixed for about 1–2 hours. This was all done in our bespoke sealed reactor to prevent the alcohol evaporating before it has had time to react.

BIODIESEL FOR TECHIES

For the technically minded, biodiesel is 'carbon neutral': the fuel produces no net output of carbon in the form of carbon dioxide. This is how it works. As the oil crop grows it absorbs CO_2 when the fuel made from the oil is combusted, the same amount of CO_2 is released. (This is not the whole story, however, as CO_2 is released in the production process. The solvent extraction of the oil, refining, drying and transporting all require energy whose production generally results in the release of greenhouse gases. In addition, if fertilizer is used to fertilize the fields in which the oil crops are grown, its production will release CO_2. To assess the impact of all these sources properly would require a thorough life-cycle analysis.)

143

SEPARATION

Once the reaction is complete we were left with a mixture of biodiesel (hooray!) and glycerol (boo!). Luckily for us the glycerol is heavier than biodiesel, so it separated out naturally and sank to the bottom. We switched off the mixer and left it for about 2 hours (ideally it should be left for about 8 hours but we were impatient). It is worth noting that most of the impurities in the mixture (such as leftover methanol and sodium hydroxide) are also taken out with the glycerol, so it's a pretty horrid cocktail. We drained off the glycerol and it came out as a gloopy black liquid. Then we drained off the much thinner, brighter, biodiesel and placed it into the wash tank.

WASHING

You may have been wondering why we dried the chip oil at the start of the process. If there was water present in the reactor during mixing instead of producing biodiesel you would make a lot of soap! Even though every measure is taken to remove water, it is inevitable that some moisture will find its way in and this results in the biodiesel containing some soap (the amount of soap also depends on the quality of the oil used and other complicated factors, but I'm afraid my 'O' level didn't stretch that far). The biodiesel also still contains some of the methanol and sodium hydroxide that the glycerol failed to catch and these could easily damage a diesel engine so must be removed before the biodiesel can be used.

Therefore, we washed the biodiesel several times with clean water. Jim found a great way to wash it using a technique known as 'bubble washing'. Water is slowly added to the biodiesel (about one third water to two thirds biodiesel) and then a fish-tank aerator is used to bubble air up through the liquids. It is basically very gentle mixing. You stop after about 5–6 hours and allow it to settle for another 5–6 hours, then the water is drained off and the process repeated until the wash water is clear, apparently up to five times – we succeeded after four. The biodiesel is dried one more time and then stored ready to be used, finally!

The final product

The final biodiesel was clear, amber-coloured, pH-neutral nectar – not bad for a first attempt. We started off with 50 litres of waste vegetable oil and finished with about 40 litres of biodiesel. Not the greatest yield, but we should be able to improve on this over time.

Costs and the dreaded tax

Get over it, the tax has to be paid. In the UK biodiesel has some preference over fossil fuel. It is currently 27.1p per litre. The only certainties are death and taxes. Therefore, it is essential that you ensure tax is paid on any biodiesel before you buy or make any. Once you make your own biodiesel you need to register with Customs and Excise.

The set-up costs were minimal as I managed to 'acquire' all the necessary containers. The breakdown on the opposite page shows our current running costs.

Biodiesel budget

It's easy to see that if we had had to pay for the vegetable oil this might not have been viable. However, the price of fossil fuel is going in only one direction: up! Undoubtedly, biodiesel has a future. It reduces carbon emissions, so its global impact on climate change is low, and it also reduces noxious pollution levels. Biodiesel has a lot going for it, but the political will does not appear to be there to take on the power of the fossil-fuel-based multinationals and promote a fuel source that could compete with oil. While we wait for the oil barons to realize it may be worthwhile growing their own fuel, it would at least be nice if there were more tax breaks to encourage entrepreneurs to see what they can do.

The only certainties are death and taxes. There is a legal obligation to pay excise on fuel. Therefore, it is essential that you ensure tax is paid on any biodiesel you buy or make

BIODIESEL: CAPITAL COSTS

Item	Cost	Comment
50 gallon storage barrel for dirty oil	£5	Scrap oil drum
Jam net to sieve the lumpy bits	£2	From the local market
Filter	£40	From Sustainable Technology Ltd
Bucket of drying beans	£100	From Sustainable Technology Ltd
Old hot-water tank	£25	
Mixing pump	Free	Cheers, Phil
Plumbing bits (valves, pipe, etc.)	£40	
Wash barrel	£8	
Fish-tank aerator	£10	For washing biodiesel
Miscellaneous funnels	£10	From the local hardware store
Latex gloves	£6	From the local hardware store
Chemical goggles	£6	Chemical suppliers
5-gallon containers	£15	
Total	£267	Could have been lot less, but our set-up looks brilliant!

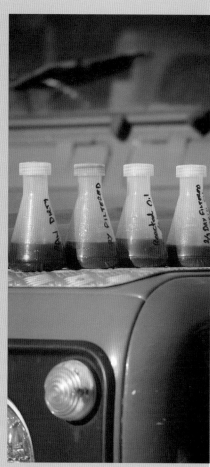

FUEL COSTS

Ingredient	Cost per litre of biodiesel	Quantity and source
Used chip fat	Free	A waste product
Methanol/ethanol	10p	45 gallons can be bought for about £100. Look for a local chemical supplier or on ebay. Smaller quantities will be more expensive
Sodium hydroxide (caustic soda)	2p	2½ pounds costs about £4. Obtain from chemical suppliers or hardware stores, but check purity (must be 99 per cent)
Tax	27.1p	www.hmrc.gov.uk Form EX403 & HO930
Total	39.1p	

Biodiesel side effects

🔥 Our exhaust smells like a barbecue and you find yourself ravenous at the most obscure times.

🔥 Biodiesel has a lower energy density than conventional diesel. Roughly speaking the reductions are in the vicinity of 10 per cent. To date we haven't noticed a great deal of difference.

🔥 Viscosity at low temperatures can make cold-weather starting difficult; biodiesel is at a disadvantage in this respect. It can be overcome by introducing a special additive or a proportion of ordinary diesel. However, low temperatures aren't a huge problem in Cornwall.

Travel light

For us as a family, our move to Cornwall threw up lots of transport issues. Cornwall is at the edge of the UK and takes a long time to get to. James was at York University and Charlotte was about to go to Newcastle; I frequently travelled for work. We arrived in Cornwall with the transport facilities on the table opposite.

To minimize the amount of fossil fuel we were using, we had to make changes. Our growing knowledge of biodiesel, steered us towards diesel vehicles, but we weren't prepared to leave performance and comfort behind. I am not yet able to think of parting with my motorbike, but I believe I can just about make a case for having it as it cuts down journey times when Cornwall is full of visitors and the roads are clogged. Somehow I managed to keep the Audi all summer; it was because of the towbar and the constant need to go and collect another trailerful of something. That said, when the Audi did go, it was surprisingly easy and we never looked back.

Changing vehicles

To use biodiesel, we had to change our vehicles. Not all vehicles react the same way. We opted for a VW as their 96–04 engines are stated as being biodiesel compatible. (Check www.biodieselfilling-stations.co.uk for information on other types of car). We started by adding 20 per cent biodiesel to our fuel; after four weeks, we were using it by itself.

As well as a VW, we wanted a Land-Rover for flexibility (load-carrying and off-road use): preferably an old and tatty one. Jim and I set out on the quest to find a budget diesel Land-Rover. Easier said than done! We travelled Cornwall for weeks and looked at any diesel, any age, anywhere, without finding what we needed. We finally realized we'd have to go to a dealer, even if it meant paying more. Enter Gary at GMO Cars Ltd in Penzance. Gary has two sales rooms very close together, the more

OUR TRANSPORT BEFORE THE MOVE

Vehicle	Comments	Prognosis
My Audi 2.6 automatic cabriolet	27 miles per gallon Lovely to drive Moderate emissions	Would have to go
Brigit's Ford Ka	40 miles per gallon Modern	Efficient, sensible little car but reliant on petrol – had to go
James's Austin A35	40 miles per gallon 1957	Reliant on petrol, but long-lived
My Triumph Thunderbird Sport motorcycle	45 miles per gallon Very useful in the Cornish traffic during summer	Reliant on petrol, but I love it
2 bicycles	Vastly underused	Perfect

expensive and lower-mileage cars are at one end (that's where we went shopping for the VW for Brigit) and the trade-ins are a couple of yards away (where I went searching). We went for the VW Polo – at just 73 miles per gallon it won out over the Golf option – and out of several Land-Rovers I took the cheapest diesel. I love driving it and it works hard; I couldn't ask for more.

There are more ways of switching to biodiesel than you may think. Before we left Malvern, we spotted an interesting vehicle at the Three Counties Showground. It was a diesel, two-wheeled, trailer-pulling all terrain vehicle (ATV). With oodles of miles to the gallon (over 120 depending on how it was driven) and a very low top speed, it would be great for short journeys to collect animal feed, remove rubbish, and so on. Unfortunately, with a very small smallholding I couldn't justify buying one. Check out www. ecorider.com – a couple more acres and I probably would have got away with it.

'Green' cars, old cars and no cars

A lot of cars on the market make a virtue of how green they are. To be fair, there have been great strides taken to reduce emissions and to increase efficiency. However, I am ever the cynic and having green vehicles without addressing what fuel they are using does not make much sense to me. It is probably better to describe them as 'greener' rather than green. I'm not suggesting anyone is lying, but some advertisements are misleading. The phrase coined to capture this phenomenon is a 'green wash'. It takes effort to scrape away the veneer and determine how green a product really is. Unfortunately, this constant need to question the efficacy of an eco-product will eventually undermine the market for genuine goods and entice sceptics to write sustainability off as a minority activity for those with more money than sense – yet another thing that pisses me off!

151

DID YOU KNOW? You could be spending about a sixth of your annual food budget – £470 – on the packaging! That's enough to buy lots of very tasty goodies, loose or in paper bags.

I like the innovative nature of hybrid vehicles that can store and use energy under computer control to achieve maximum efficiency. That said, I would like to fully understand the life expectancy of the electric energy cells and have yet to look into the expected life-cycle energy expenditure of such cars. I wonder how well they will perform against James's Austin A35?

Old verses new

Brigit had always had old cars – Morris Minors, Fiat 500s, Ford Populars, and so on – and I think some of the love for old machines rubbed off on James. He decided he wanted an old car that had all the production energy well and truly sunk in it, yet still had a lot of life left. Just before we moved he bought an old 1957 A35. There is an argument for buying and keeping an old, reliable car. As a rule of thumb, the average car produces as much pollution (there is a direct relationship between energy used and pollution created) in 9 years of running as is produced to manufacture it. If it is assumed that the average new vehicle has an average life of 15 years before it is effectively landfill (sad, but true), then keeping a car that is capable of longevity can, over time, save a lot of energy.

There is more to travel than cars

Bicycles are great – we should do more cycling. Walking is great – we should walk more. I think we all know what we should be doing, but sometimes it just isn't possible. Too much shopping, too little time, too far, bad weather – person-powered transport is not always palatable. We do walk to the shops, butcher's, post office and pub (OK, the pub is hardly a chore) in the village, but that's because they are close and convenient. We don't use our bikes often enough because the hills, although small, put us off. We will try to do more.

HOW THREE NEW CARS CREATE MORE POLLUTION THAN ONE OLDER CAR

Years	One old car	Three new cars		
		Car 1	Car 2	Car 3
0	1 unit of pollution to manufacture	1 unit of pollution to manufacture		
3				
6				
9	1 unit of pollution to run	1 unit of pollution to run		
12				
15		$^2/_3$ unit of pollution to run	1 unit of pollution to manufacture	
18	1 unit of pollution to run			
21				
24			1 unit of pollution to run	
27	1 unit of pollution to run			
30			$^2/_3$ unit of pollution to run	1 unit of pollution to manufacture
33				
36	1 unit of pollution to run			
39				1 unit of pollution to run
42				
45	1 unit of pollution to run			$^2/_3$ unit of pollution to run
	Total: 6 units of pollution	Total: 8 units of pollution		

Before we moved to Cornwall Jim sold his narrow boat. Bless him, he did without a boat for about three weeks before he found a very tired little Shetland boat that was exactly what we needed to go fishing when we arrived at the seaside. Unfortunately, it needed an awful lot of 'tlc' and the old engine was a goner (actually, this was probably fortunate, as if it had been working there was a good chance it would have been pumping out a serious amount of pollution). In the spirit of our endeavour to be green Jim knew he had to find the most eco-friendly way of sorting it out. Initial investigations uncovered a diesel outboard that we thought we could run off biodiesel, sadly, the demand was not there and it proved impossible to buy one, so Jim set off on a quest. As I've already pointed out, Jim spends a lot of time researching on the internet. He did himself proud with the amount of time he put into finding the right solution for his boat. He was like a cat that had just got the cream when he finally tracked down Mariner engines with ultra-low emissions. Matt at Barrus in Oxfordshire came up trumps (www.barrus.co.uk).

Having sorted out the power it was a mere matter of making it float and steer properly. Just up the road at Lostwithiel, Jim found Outboard Services and Marine (www.outboardservices.co.uk) who knew all about old Shetlands, so he managed to sort out the necessary goodies for the steering.

After a bit of trial and error (Jim can tell you about the leak!) he and his step-dad, Maurice, who only lived down the road in Truro, managed to get the boat to stay afloat. We were ready to fish and generally enjoy a life afloat the low impact way. I had a chance to go sailing with James and we had a brilliant time, but we have not yet made the most of the boat, time and weather have been against us. Having said that, when we get a chance we're off!

We are also very aware that public transport is there to be used (even if it does rely on fossil fuels). We are half a mile from a station with direct trains into London. We try to use the trains, and encourage our visitors to use them, as much as possible, but somehow trains never quite seem to

AIRLINE TRAVEL

The impact of all our other transport decisions is totally eclipsed by our desire and our need (yes, need) to fly. I was lucky enough to make two trips to India and South-east Asia for television programmes about the Second World War. I was playing a part in telling the stories of those veterans involved in the 'Forgotten War' campaigns in Singapore, Burma, Thailand, Malaysia and India. It's true my flights will have caused serious pollution, but I feel I can justify them. And it's not just me. Brigit has travelled and experienced different cultures and places. James and Charlotte are young and cannot wait to see the world. James had a gap year in Nepal, loved it, learnt a great deal and grew as a person. Charlotte had a gap year in Cuba, loved it and has a taste for exploring.
The world is small if you fly, and very large if you try to cross it by land. So how can you reconcile the desire to travel and the desire to be green? We haven't got to the bottom of it yet. We have planted trees specifically in answer to those long-haul flights. It's better than nothing – just. In future, I believe we will consider the impact of our travel arrangements before deciding on a holiday destination that we have to fly to – there are hard decisions ahead.

travel when you want them and, if a couple of people (or more) are travelling, it all gets rather expensive. We are also very lucky in that there is a proper bus service here. In Malvern there was about one bus going in each direction per day. Now we have an hourly service running until midnight. We always consider public transport for longer journeys but less often for the short- to mid-length ones where convenience is the primary motivator. Other forms of transport can always be considered. When judging a series of Scrapheap Rally races, I was introduced to the concept of an electric rickshaw by one of the teams who built them for a living – an awesome idea; not cheap but great fun.

Check out www.cyclesmaximus.com.

Heating the house from hell

New House Farm is a lot bigger than we had intended to buy. In many ways that's great and we have fun with all our visitors. The downside was that it was draughty, damp and generally cold, which was not good with the winter of 2005/2006 threatening to be the coldest for ages. We needed to ensure we warmed it and kept it warm. We did masses of research before finally deciding what sort of heating to use and which type of insulation would be best.

The house is made of a mixture of stone, brick and cob, with bits of it rendered in cement, which doesn't breathe, and other bits rendered in lime or painted. The only thing that was consistent was the pervading damp. It was possible to sit on a sofa or chair only at certain times during the summer and not get up with a wet botty – not conducive to healthy, comfortable living. The damp was probably a bigger problem than the cold, but anyway we had both.

It wasn't until October that we could spare enough time to sort out our heating system. We had plenty of advice and much of it was contradictory,, so decision-making was not easy. But I've always felt that the best thing about advice is that it's free and you can ignore it.

What we couldn't do

We knew we did not wish to use fossil fuels. The fact that the building is listed, and the archaeological interest in the land, made a ground-source heat pump impractical: it involves laying a pipe underground to extract the heat from the earth that is stored when the sun's energy is captured by the soil. Typically you can put 1kW in and get 4 out, which is not bad, but the heat (not hot but quite warm and a lot of it) is best suited to something like underfloor heating. Our ceilings were so low we couldn't contemplate shaving off a couple of inches to accommodate the pipes, even if the building hadn't been listed.

What we could do

We had a wood-burning stove in Malvern and loved it, so we wanted something similar in Cornwall. Typically, life isn't that simple. There are lots of stoves on the market, all with pros and cons. We considered wood, multi-fuel, wood-shaving, sawdust and wood-pellet stoves. Apparently, wood pellets are the new gas, but we didn't like the amount of processing involved in making them. Sawdust and wood shavings would be great if we had a source of them as waste products, but we don't, and the 'multi' part of multi-fuel tends to have fossil origins, so burning good old-fashioned logs appears to be the most sustainable.

Apart from the small amount of energy

It takes a lot of effort to keep the woodburners fed, but they work well and look great

Fuel and Travelling

expended in cutting the logs, during the tree's life it will have provided twenty times more oxygen than is required for combustion, and will in turn release no more CO_2 than would eventually be released when it decayed. And there is the added advantage of real flames.

Choosing a wood-burning stove

Brigit had spent a lot of time researching wood-burners before we had bought the Clearview stove we had in Worcestershire, so, after a brief trawl to see if any more pretenders for the crown of most efficient

and lovely stove had entered the arena, we contacted Helen and Jonathon at Clearview again (www.clearviewstoves.com). Helen and Jonathon know wood-burning stoves. They make them in their own factory out of the best material and they are not happy to sell until they know you have exactly the right stove for your needs. They had to help us chose four for the main heat sources for our chilly old house!

We opted for a Solution 500 with back boiler and a flat top for cooking on for the kitchen, a flat-topped Vision for the hall, again with a back boiler, a Vision with a low canopy for the sitting room, and a little Pioneer for the room Brigit and I were sleeping in that would eventually become a craft room/study. In one fell swoop we had all the downstairs heated to perfection. Clearview stoves are seriously efficient and extremely well made. With a very little maintenance ours should outlive me with no difficulties. If you keen to have a wood-burner, they are definitely worth looking at.

One thorny issue has to be addressed: the wood. If you don't have good seasoned wood, you are wasting your time – the heat output is way down and the stoves difficult to keep lit. Freshly cut, 'green' wood has a very high moisture content, but if it is allowed to spend seasons stacked out of the rain, the amount of water in it reduces massively. Our first batch of wood came from Helen and Jonathon and it caught

159

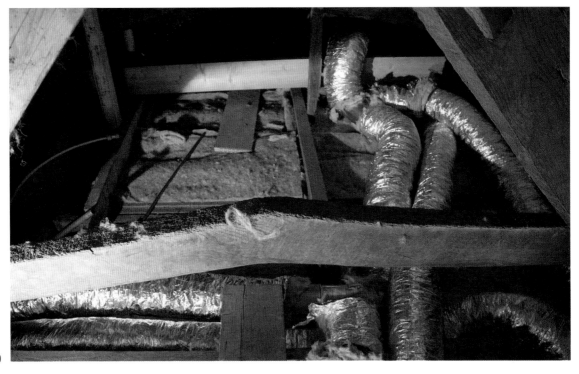

after little more than having a match held under it, thanks to years of seasoning and being kept dry. Our second source was not well seasoned, still damp and impossible to light. A chat with our local chimney engineer, Steve, and we found Ken, who seasons his wood after it has been split and keeps it dry – brilliant! We now have a decent-size wood store that is sufficient to keep us for a year or two and gives us time to go and find our own wood. It appears that getting green wet wood is easy but you have to be prepared to put the necessary work in collecting, splitting, stacking and drying – thank goodness for the children!

Distributing the heat

Having found sources of heat, it was time to work out how to spread it around. First and foremost it makes sense to move the air around the wood-burners so it circulates. Enter the Caframo Ecofan – a heat-powered fan designed specifically to circulate the warm air created by a stove. This fan does not use any batteries or mains electricity and it's not magic. It has a thermoelectric module which acts as a small generator to power its motor. When this generator experiences a heat differential between its top and bottom surfaces, it pumps out electricity. The bottom surface of the module is heated by the base of the fan sitting on the stove, while the top of the module is kept cooler by the fan's top cooling fins. Second, we had back boilers which could have been used as part of a conventional heating system, but Chris Hendra, a knowledgeable and forward-thinking 'green' architect, introduced us to the idea of whole-house ventilation and we liked what we heard. A search on the inter-net and several phone calls later and we found Mike at Ubbink (www.ubink.co.uk). Mike has a vision that one day every house

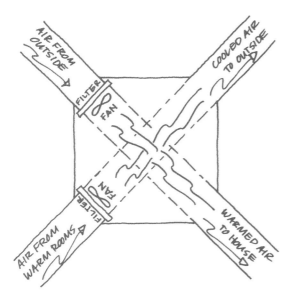

in Britain will be a Ubbink house. He is either going for world domination or truly believes in the system (it's the latter). The idea is very simple. Your house is made up of warm areas, warm wet areas and cooler areas. All you have to do is get rid of the moisture, and take some of the heat from the warm bits and pass it to the cold bits. What is needed is a set of ducts connecting the extractor vents from the warm areas to a heat-recovery unit, and then another set supplying warmed fresh air to the cooler areas.

In practice, it was fairly painless. Mike, Martin, John and Steve turned up, Jim and I pitched in, and a day and a half later the roof space looked like a scene from *ET* with lots of insulated silver ducts from 4 to 8 inches in diameter connecting all the rooms' vents to the heart of the system (well, more the lungs than the heart), which is the heat-recovery plant. The plant is low-energy – about the same as a couple of low-energy light bulbs – and can be set to different degrees of air distribution. The air in the house changes about once an hour, and the spin-off is that the system filters out the allergens, dust mites and so on, so the air is a lot cleaner and healthier.

The functions of the New House Farm heat-recovery unit

🔥 Extract: suck warm air from the same room as the wood-burners (not right beside them as we want the warm air to circulate first); suck warm wet air from the bathrooms.
🔥 Condense moisture in the extracted air.
🔥 Recover the heat from the extracted air.
🔥 Filter and vent the spent air to outside.
🔥 Bring fresh, filtered air into the house.
🔥 Warm the air coming into the house from the outside using the recovered heat. The incoming air crosses over the warm air that has been extracted from warm parts of the house in a number of parallel overlaid capillaries. This heat transfer is dynamic and allows higher efficiencies to be achieved.
🔥 Supply warm clean air to the cooler rooms (the upstairs bedrooms).

At first we successfully took all the heat out of our downstairs rooms and threw it up into open loft doors and holes in the ceilings and walls, but it didn't take long before we managed to close down the escape routes. Now we are gradually drying out years of neglect – and it makes a huge difference.

161

Fuel and Travelling

Insulation

A key element in keeping our heat in was insulating the roof space. Brigit conducted the research. Our criteria were these: the insulation had to be effective, sustainable/eco-friendly and, if possible, cheap. There is quite a lot out there and we had to read several of the labels more than once to understand fully the nuances of how environmentally friendly they were – or were not. In the end we selected and used two types of insulation.

In the newer part of the house (only 150 years old!) we used Innotherm, which is largely made from recycled cotton and denim with some polyester to bind the fibres. It was being made in America and is now produced in Yorkshire (www.recovery-insulation.co.uk). In the larger older part of the house we used Thermafleece, a British

wool insulation that is suitable for a variety of roof, wall and flooring applications. Brigit particularly liked the fact that this product found something useful to do with fleeces that otherwise had little commercial value. Second Nature UK (www.secondnatureuk.com) uses the coloured, coarser wool from the sheep that graze on the upland hills throughout Britain (breeds such as Swaledale, Herdwick, Welsh Mountain and Scotch Blackface) to create an extremely efficient and natural insulating material. It is not the cheapest but it is very easy to use and something about it tickled our fancy. A woollen coat for the house! The website has lots of useful information, including local stockists, but what is a must-read for all the eco-engineers out there is the technical data page. It makes you look at a sheep in a very different way.

OUR TOTAL FUEL USAGE

Use	Fuel Solution	Comment
Car 1	Land-Rover – biodiesel	Green and effective
Car 2	VW Polo – biodiesel	Green and effective
Car 3	Austin A35 – petrol but durable	Justifiable (as far as any family having three cars can be justified)
Motorbike	Petrol	Not good other than for my sanity – sorry
Insulation	Textiles and fleeces	Green
Heating	Wood-burners and whole-house ventilation	Green and effective
Cooking	Electricity and LPG gas	Not green but very capable*
Electricity	Ecotricity Solar PV Waterwheel Wind turbines	Greener than a green thing wearing green hemp clothing!

*Food is a subject dear to my heart. Choosing our means of cooking was not a trivial matter. All our discussions about ranges and cookers kept coming back to the fact that we liked cooking with gas and we liked the quality of the Mercury we'd had in Malvern. It may sound lame, but we couldn't find anything we wanted more, so we bought another one. The electricity is covered by our green tariff and home generation, but the gas is an issue for the time being, though only until I'm allowed to generate my own!

WHEATIES

When we have guests that feel the cold Brigit is always there to provide them with a 'wheaty'. Like hot water bottles, wheat or rice bags are great for keeping warm without turning up the heating (if you have any). Simply put the bag in the microwave for a couple of minutes and it will keep warm for up to an hour. Ideal for taking to bed when chilly, or for warming you up at any time. Also great for treating aches (including earache, tension, tummy pain, etc.). Wheaties are now widely available in the shops or you can make your own – take a clean sock, fill the foot bit three-quarters full with rice and heat it up in the microwave. Make sure you switch off your microwave at the wall after use to save energy.

Our way forward

I don't think we did too badly for our first winter. The sheer size of our house makes it difficult for us to warm it to the preferred temperature of some of the people who live/stay with us (you know who you are!) but, all in all, it is comfortable and the heating is effective. We are now comfortable enough to survive, but there is still a lot more we have to investigate.

Ever since Tom and Barbara made their own gas-powered generator in *The Good Life*, I've wanted to make my own biogas system. As if that were not explosive

enough, I've yet to attempt hydrogen cells or building a gasificator. Watch this space …

Brigit and I visited Chris and Gilly Hendra at their very green, newly built house. It was lovely and they had the most amazing two-storey conservatory that helped warm the fabric of the house and produced lots of warm air to feed into the whole-house ventilation. We are chatting to the planners and Chris is coming up with a design.

There is definitely more potential for micropower systems at New House Farm. A micropower system is just a generic term for a small domestic generating system which can be powered by wind, sun or water.

Blowing Hot and Cold

Energy and the weather

Most of us have contemplated generating our own electricity (usually when we get our quarterly bill) but have been put off by the complexity and/or the expense this involves. We are now fortunate enough to have simple, clean, renewable energy systems that allow individuals to provide themselves with sustainable electricity and thereby contribute to a reduction in the greenhouse gases emitted in our names. Waterwheels and windmills are part of our heritage and we are aware of the advantages our ancestors obtained hundreds of years ago by harnessing the energy that drives them. Not everyone has a source of water in close proximity to their home, but Britain definitely has lots of wind – probably the most developed and easily understood of the energy sources. We are an island and the wind often feels as though it has made its journey across the Atlantic just to buffet us.

Cornish weather

We decided to live by the seaside, so of course we knew there would be no shortage of wind power. When we decided to buy New House Farm this was one of the plus points. Our valley runs to the south-west, and everyone knows the prevailing winds in the UK are south-westerlies. We knew all we had to do was tap into the free resource that would roar up the valley and generate as much power as we wanted. However, things were not as straightforward as we thought they would be.

The Cornish weather is far from predictable. When we moved in to the house we were given lots of useful information about the climate – it's windy, it rains a lot, it never snows and frosts are a rarity – but that is not exactly what we experienced in our first year. It was windy, but often from the north – over the village – and much disrupted before it reached our little part of the county. For the first three weeks after we put up the turbine to power the pump from the well there was hardly any wind at all. In fact, when it was windy enough for us to see the blades go round it became an event worth stopping work for. It did rain a lot during summer when we didn't have a roof but, surprisingly, after we had spent a fortune sorting out the leaks, we had quite a dry autumn and winter. Thank goodness our waterwheel is fed from a spring!

We regularly had beautiful, crisp, clear days that resulted in beautiful, crisp, clear nights and frosts down to –6/7°C. The combination of our newly built wall and the greenhouse heating system functioned better than we could ever have hoped, and

the winter salad plants remained frost-free (though I think we had more than our fair share of slugs that took refuge in and around the raised beds). One weekend, in November, we had a very pretty 5-inch snowfall and on several other occasions we experienced sleet showers that settled as snow on the hill around us. When we took the dog for a walk on the beach she played in several inches of snow all the way down to the waterline. It makes you wonder how the palm trees survive down here.

I really like the weather we get thrown at us in Britain, so it was lovely to experience the extremes as well as the temperate periods we expected. One thing we learnt very quickly: it pays to ensure you don't bank on any one particular weather phenomenon acting as an energy source.

The wind farm debate

Driving through Cornwall on the A30 you see evidence that lots of wind power is available. Large turbines are not common in the UK but they do exist, and the most southerly section of the road brings you in close proximity to several wind farms. Some are right beside the A30. Not everyone likes the look of these large, white structures. Some people love them (Jim thinks they are awesome as well as great examples of engineering) and others hate them for spoiling the natural, beautiful scenery. I fall into a third camp. I think they are necessary, not exactly ugly – although I prefer the countryside without them – and very functional.

We arrived in Cornwall to a heated debate about a proposed wind farm in the valley next to the one where we were going to live. I call it a 'debate', but a proper debate (in which the reasons for and against a proposition or proposal are discussed) requires people to listen as well as talk. If you want to see raw emotions on display, put in a planning application for a large turbine and stand back. Protests are still being considered, although our local council and the protesters are both on record as saying they are in favour of individuals producing their own electricity on a micro-level, and that it is just the larger-scale projects they have issues with.

I must admit, I find it frustrating that any proposal for sustainable energy is judged on a case-by-case basis and not in the context of a national strategy that has emerged from grown-up, informed discussion. Every location is bound to be in

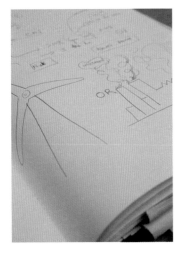

POWER AND POLITICIANS

Sound bites suggest that most politicians are in favour of sustainable power, but there has been little real action. In fact, the only committed stance I was aware of in the latter part of 2005 was the government airing nuclear energy as the panacea for our long-term energy needs. And, despite the argument that it is clean, cheap energy, I am too cynical to be happy with a 'for profit' organization looking after something as dangerous as nuclear waste. I just can't get my head around what is involved in storing something with the half-life of nuclear waste, and the management this will require. At some stage it will all go horribly wrong, and the ball will be passed back to the good old British public.

Blowing Hot and Cold

someone's 'backyard' and all that happens when yet another proposal is tabled is that more antipathy is generated. Perhaps we need to give regions or counties, or even districts, the responsibility to provide their solutions to sustainable energy. 'Not here' isn't an option; at the least we need to agree there is a problem.

Harnessing nature to generate electricity has to be the way forward, be it the wind, the sun or water (sea and/or fresh). We have to get engaged, as individuals and at a commercial level. On a grander scale there are many exciting initiatives that will provide truly green energy. For example, a company at the Scottish Green Energy Awards in Edinburgh in December 2005 had pioneered offshore turbine erection with techniques similar to those used in the oil and gas industry. Offshore wind will undoubtedly soon become more exploitable and must surely be the best of all possible worlds.

No one can complain about the aesthetics of something that is very hard to see and there is a plentiful supply of wind out at sea. You'll be glad to hear that I've just carried my soapbox out to the barn and promise to leave it there!

Wind facts

✶ Wind is abundant around the UK.
✶ Commercial wind turbines don't take long to produce energy equivalent to that used in their production (which also means they pay for themselves quickly).
✶ Electricity generated by the wind is sustainable.
✶ Hill tops are the best sites for reliable wind; unfortunately, the turbines stick out like a greyhound's dangly bits.
✶ Small-scale turbines are available and affordable.

Where does wind energy come from?

Wind energy, like solar, water and even fossil-fuel energy, ultimately comes from the sun which constantly radiates about 175 thousand million megawatts of energy to the Earth. About 1 per cent of this is converted into wind energy (that's about a hundred times more than the energy converted into biomass by all the plants on Earth).

172

The sun's energy makes the land hot. This heats the air, which becomes lighter and rises, and is replaced by the adjacent cold air. Wind! But it's not all about air moving from A to B in a straight line. The irregular shape of the land masses, the rotation of the Earth and a bit of the Coriolis effect (that's for the techies) and you end up with some pretty complicated global winds that determine the prevailing wind direction. As if this weren't enough, things affect conditions locally – sea breezes and mountain winds, for example, have significant impact. The bottom line is that winds are complicated and always changing. Luckily there are people who spend their lives monitoring and recording what happens, so if you want to put up a wind turbine it is possible to get information about what to expect in your area without having to understand the entire process. Try the British Wind Energy Association (www.bwea.com) and follow the step-by-step guide.

Wind power follows the laws of physics, which keeps things nice and simple. The cold, cruel fact is that little turbines make less electricity than big ones, which is not surprising as it is all about how much air is caught on the blades. That said, many blades aren't more efficient than a few well-designed ones. The metal pylons with lots of blades that are synonymous with isolated ranches in America or cattle stations in the Australian outback tend to be pumps for water. Pumps are slow and steady, but alternators to make electricity like fewer, faster blades. Obviously, the efficiency of the blades is important, as is the efficiency of the alternator you have attached to them. You don't get anything for free in this world, and you tend to pay for better performance.

The amount of energy you can hope to harness depends on where you live. People in Scotland, Wales and south-west England can expect about 3000 hours of maximum operation per year. With a 1kW (1000W) domestic system this is the same as upwards of 3000KWh per year. On a green tariff the charge is about 12–14p per KWh, so you get about £400 worth of electricity per year. From there it's simple enough to do the sums and work out how long it will take to pay back your outlay. Notwithstanding the economics, using wind power provides protection against the energy price rises that will undoubtedly occur in the near future.

Building our own wind power

Our first wind turbine was cheap and cheerful. We didn't need to push for too much power as it was needed only to charge a battery for a little 60W demand pump for the well that we reckoned would use about 30Wh a day (over a 24-hour period it would be on for a total of about 30 minutes). Ivan at Navitron came up trumps, and at just under £300 for a complete system (mast, blades, alternator, and charge controller complete with dump load) the turbine was a bargain. After a couple of months, we customized our very cheap-and-cheerful square-ended blades to reduce the noise. I had reckoned that if the noisy blades generated complaints we could do something about it, but if we did the modification straight away there was nowhere for us to go to appease the neighbours. As it was, the noise wasn't too bad and we had absolutely no complaints. However, it can sometimes be a bit *Hound of the Baskervilles*.

Turbine 1 was near the well, for convenience and to make full use of the wind coming up the valley. Turbines like non-turbulent air passing over the blades so they should be sited to minimize the influence of obstacles (ideally this should apply to all directions, but our valley location dictated that we had issues with wind from the north). As a general rule of thumb, the distance between your mast and the nearest obstacle should be ten times the height of the obstacle – for example, a turbine needs to be 33 yards away from a 3-yard hedge. It is stating the obvious, but the mast must be sited so that the blades catch as much wind as possible – the higher, the better.

It is essential to look at the power curves that show what you can expect to achieve at what wind speed. And the turbines must bc able to protect themselves when the speed is very high, otherwise they could spin to destruction or burn out if they try to produce more electricity than they are meant to. Most have a simple system that involves the tail causing the blades to turn a bit away from the wind when it is very high so that they catch less of it and slow down. This is called furling. We have had several evenings when the wind has been ferocious and we can hear the blades turning out of it. I love it, and the noise I hear is free electricity being made!

173

Turbine construction

During the summer, our slave labourers (James and friends) had dug a hole for the foundations (not frightfully big – about 1 yard × 1 yard and about 24 inches deep – we weren't sure if this would be the final position for the turbine). I had filled it with some of the many stones the pigs had turned over while they were rooting and clearing their enclosure. A couple of half an inch threaded bars and several wheelbarrows of concrete later and I had a base plate.

Jim and I, with a bit of help from Brigit, assembled the equipment that had arrived in a few boxes and in a couple of hours we headed down to the lower paddock. Thank goodness it wasn't windy. A 4-yard mast doesn't sound high, but with the turbine on the top of it, it was quite a handful. I mentioned that we had a little help from Brigit. Well, she was very interested in the bits we were putting together and had no end of useful suggestions but as we struggled to keep the mast vertical she really came into her own with comments like: 'Are you sure that will hold it up?'; 'Wouldn't it be better to …?'; 'That doesn't look right'; 'Does it matter if the guy ropes are that way around?'; 'I don't like the red nose on it. Do they have any different colours?'. Suffice to say, it was a great team effort and, touch wood, the structure has stayed up and produced power with very few problems.

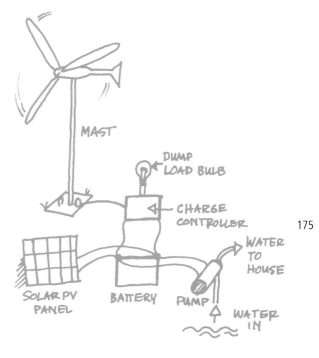

OUR WIND POWER SYSTEM

Turbine 1 – Navitron 200W 24V system (provides about 140W at 12V)

Mast

Charge controller

Blades and turbine

Stakes/guys and base plate

60W polycrystalline solar photovoltaic pane for back-up if there was no wind

Jim's homemade charge controller and rectifier (for 12V) including sexy little box with lights

Dump load made from an old 12V car main-beam light bulb (the dip element was broken – REUSE!)

110-Ah deep cycle battery

12V Flojet pump

The noise of the turbines is the sound of free electricity being made

THE THEORY

If you want to go into some of the theory, the power available from the wind is proportional to the cube of the wind speed and to the square of the blade diameter. Therefore:

• A 20 per cent increase in wind speed increases the power available by 73 per cent.

• A 20 per cent increase in the blade diameter increases the power available by 44 per cent.

Some problems

There were a few teething problems. The stakes provided were not as secure as we would have liked so we drove in three lengths of 1-inch box steel to give extra support. We had an instance of the electrical wires coming undone and Jim didn't like not knowing how much charge we had in the battery, so he built his own charge controller from scratch. It has a green light if the battery is fully charged and a red one if it is charging. To begin with, lack of wind meant there were several occasions when the battery ran out and we had no water other than from the mains tap in the kitchen. The first time this happened it was easily rectified. We changed the battery and gave a couple of pipes in the attic a shake to get rid of the air locks – and hey presto!

The second time, we found out only when someone couldn't flush the loo. The immediate action was the same – a quick battery change and giving the pipes in the attic a shake. However, flushing loos were definitely expected by the family as part of their 21st-century lifestyle, so we had to find a fail-safe. It was quite easy. We reconnected the mains to the water-storage tank in the attic, on a ball cock, but this time it was just a few inches up from the bottom of the tank. The theory is simple. If there the well system fails or there is a problem with Turbine 1, the tank will never be quite empty as the ball cock activates the mains when the water level gets down to the last couple of inches, and we use mains water until we can fix the problem. The system works. We know it does, because after another period without wind we checked and, although we found a flat battery, there hadn't been a disruption in the water service.

A bit of extra help

It was apparent that we could not rely on wind alone to provide energy for our pump. We have yet to exploit the potential of our stream as it crosses the lower paddock, but to do so is a fairly large job so we discounted this in the short term.

Having found a 60W solar panel on the internet for £200 and knowing that there will undoubtedly be more wind-free periods, (we have yet to use the system in summer), we decided to harness both wind and sun,

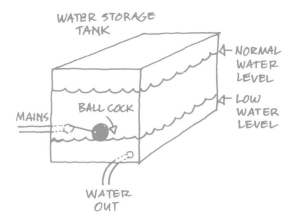

which means we now have fully automated, free water that can be relied upon.

Turbine 2

We knew we wanted more wind power to offsest what we were using from the grid and a trip to Scotland turned up trumps when I met David Gordon, the founder of Windsave (www.windsave.com). He is aiming to provide mass-produced wind turbines that can be fitted onto most properties and need very little installation. The system is very simple, it is rated at 1000W and it has a blade diameter of 1.9 yards.

The whole system is in modular form and it harnesses wind energy, makes electricity and directly wires a 13-amp plug into the mains systems (we chose to put our power on to a fused spur). When the wind blows, and you are using energy, you reduce your electricity bills. The net effect is a potential saving of up to 30 per cent overall against the UK average 'domestic' electricity bill.

The turbine details are caveated by the statement that 'the generator only provides savings when the wind is blowing strong enough to generate and when the customer is 'using a reasonable normal demand from the ring mains' (using electrical appliances). However, we have to face facts: we run fridges, freezers, telephones, answer machines and some computer equipment on standby all the time. Whatever time of day we generate power we will be using it as well so we will have something to offset it against.

Planning permission

It wasn't so long ago that the only way to own a wind turbine without paying a ridiculous amount for it was to build it yourself. Jim still has aspirations to make one, from carving the blades to winding his own alternator. Initially, we were just too busy to allow ourselves the luxury of taking time out to build from scratch, so we searched the internet for our bits. There are courses where you can learn to build a turbine for pennies – providing you don't take labour into account!

Virtually all wind turbines need planning permission and we were permitted to have only 3-yard structures within the domestic curtilage of our plot. Once we have determined the best location for our turbines we will formally apply for permission and make the installations permanent. In general, wind turbines, whether in groups or singly, will be permitted if you can answer as many questions as possible before they are asked. The checklist below is a good starting point, but it is worth remembering that even if you are not compliant on some of the points it may be possible to argue your case.

As a general rule, the more visible a turbine is, the more controversial it will be, although this will not necessarily mean that planning permission will be refused.

Also consider whether the turbine will cause a nuisance to any of your neighbours. A good rule of thumb is to put yourself in their shoes, and consider whether you would find the turbine a nuisance. Its impact on residential amenities will be assessed as part of the planning application process, whether or not neighbours make a formal objection.

PLANNING APPLICATION: FORMS AND FEES
You will need a set of five application forms, which are available from development services at your local district council, a map of the site showing the position of the turbine mast and elevational drawings showing the appearance of the turbine. These are sometimes available from manufacturers. A drawing suitable for inclusion with a planning application for a Proven WT2500 is available on the Proven website. There is a fee so you should look into this beforehand. Applications take around eight weeks to be determined and you will need to allow for this when you book contractors. It is inadvisable to start work before planning permission has been formally granted.

Planning permission checklist

* Does the turbine break the skyline? Is it seen against a backdrop of hills, trees or buildings, or the sky?
* Can the turbine be seen over long distances or from important local landmarks, especially if it is seen against the sky?
* Is the turbine a man-made feature in an otherwise unbuilt landscape?
* Does the turbine look very large in relation to its surrounding features?
* Can wires and equipment be installed unobtrusively? That is, can the wires be buried and can the equipment be housed in existing buildings?
* Can it be seen from main living areas and does it dominate the view?
* Can it be heard from inside a neighbour's house and is the noise intrusive?
* Will it cause shadow flicker across windows to the main living areas or on a significant part of the garden?
* Will it interfere with TV reception?
* Will it be very close to a neighbour's house or positioned directly above neighbouring properties if you live on a hill?

Sunshine

There are ways of capturing the sun's energy directly and we do this in two ways. The first, and simplest, is by using it to heat water – solar thermal – and the second is by using photovoltaic cells to produce electricity.

Solar thermal

Plumbing was somewhat lacking when we moved into the house, and our showers were provided either by using hot water from the Burco boiler or by our very rudimentary solar shower. This cost only £5.99 from a camping shop and consists of a heavy-duty, black plastic bag with a nozzle. The principle is simple. Fill the bag with water. Leave it on a flat surface in the sunshine and, as if by magic, a couple of hours later you have warm (sometimes very hot) water. It works, and it kept the team from getting smelly. Even when the shower and bath inside the house were functioning, several diehards preferred to use the free hot water outside. This could be for a couple of reasons, but the most likely is that there was a bit of a naturist thing going on.

Our first foray into solar thermal was an experiment by Jim and his brother Andy. We'd seen pictures of black radiators being used to capture the sun's heat, so I procured a damaged radiator from the Plumb Centre. We already had some plywood, corrugated clear plastic, a dustbin and a collection of plumbing fittings. The experiment turned into a rather Rolls-Royce solution but Jim and Andy had fun, so who am I to complain?

Jim and Andy's system had some disadvantages. The inside of the radiator gets rusty if it isn't flushed through every day. After a couple of days with no sunshine the water had a serious orange tinge. The dustbin has to be refilled after you have used all the hot water you want. Thermal siphoning needs to have a complete circuit, so the water level at the top of the bin must be above the output at the top of the radiator. Probably the biggest disadvantage was that the water was warm, but never very hot even in glorious sunshine.

Overall, Jim and Andy's experiment had to be classed as a success, but I have to admit that the solar shower bag was much easier to use.

Within the image (hand-drawn labels): WATER IN · HOT · HOT WATER · WHEN ALL THE WATER IS HOT TAKE A SHOWER! · WATER GETS HOTTER · COLD · COLD WATER

Jim and Andy's simple homemade solar heating system

* Paint the radiator matt black.
* Build a box for the radiator. You can put insulation in the back to lessen heat loss; they didn't bother.
* Put the radiator in the box.
* The cold-water input goes to the base of the radiator.
* The warm water comes out from the top of the radiator.
* Drill a couple of holes in your water tank, one at the top and one at the bottom. We used a cheap black plastic dustbin – recycled of course!
* Seal the lid of the radiator box with clear plastic or glass. This stops the air circulating and reduces heat loss.
* Place the box to catch as much sunshine as possible (south-facing and so that the sun hits it at an angle of 90 degrees).
* Fill the tank (and the radiator) with water.
* Wait. No pumps needed: thermal siphoning will move the water around.
* Use the hot water. We turned the disused byre into a very nice *al fresco* shower.

A long-lived system

We decided that we wanted a proper solar thermal system to heat the water in the house. We were reliant on electrical immersion heaters and even though we were on the green tariff it was neither efficient nor cheap. The listed status of the house made fitting any system to the main house a bit of a problem. Thank goodness we had a south-facing potting shed, home of the hens, with a rather manky tin roof but sufficient integrity to support our water heaters.

Obviously, before we started we did our obligatory research. We were keen to build our own system, using flattened copper pipes soldered on to a sheet of copper (an old hot-water tank that had been cut up and unrolled). We were about to commit to this when we found some rather sexy evacuated tubes.

A bit of physics. We know that matt-black bodies absorb heat efficiently, which is good. Unfortunately, due to reciprocity, matt-black bodies also emit heat very efficiently, which in this case is bad. So our ideal was a system that absorbs heat but does not emit it – we needed a one-way barrier. Bring on down the evacuated tube, which minimizes radiated and conducted heat losses. We found a very effective and cheap system on the Navitron website (www.navitron.org.uk). As valued customers, we even persuaded Ivan to come and play with us so that we could pick his brains about the installation. We had a great couple of days putting in our very own, very efficient, water-heating system. It was just a pity that it was December and not exactly a prime solar-thermal season.

We spent a day installing and a day commissioning and diddling. We have had lots of guests since we arrived in Cornwall and expect a lot more in the future, so we decided to ensure we produced sufficient warm water for the masses. Everything suddenly got rather big.

Catching solar energy

✹ Shiny/light surfaces reflect the heat – white and silver cars are cooler in the summer.
✹ Black/matt surfaces absorb the heat – black cars may look cool, but they aren't.

OUR SOLAR-THERMAL SYSTEM

Product	Cost	Comments
Two lots of 20-tube solar panels	£790	Twice as much as would be needed for an average house
Panel mounting straps	£10	
50 yard roll of twin stainless-steel insulated pipes including control wires	£495	A really worthwhile expenditure. It was a long and complicated route from the potting shed to the attic. Using copper would have been a nightmare because of the twists and turns and plastic couldn't take the temperatures generated by our system
Pipe fittings	£48	
High temperature insulation	£14	You can't afford to lose any of your hard-captured heat
New solar water cylinder	£385	Really clever, tall and thin, which reduces convection losses. Two indirect feeds, one for the solar thermal (at the bottom) and one that we could use for the wood-burners' back boiler. Temperature sensors can be inserted top and bottom. Holds enough for a big bath for Brigit as well as some showers, double insulated, so hot water can be saved for as long as possible
BS3 solar controller	£130	This little gizmo gives the temperature of the hot water at the top of the tank and the cold (input) at the bottom, and the temperature of the 'collector', that is the water being heated outside. It then decides when to use the solar-heated water and controls the pump – oh, and it logs how many hours it has been active. I like it!
Central heating pump	£45	
Small expansion tank in the loft	£15	
Miscellaneous copper pipes and bends	£10	
Solar antifreeze	£44	Must be safe for human consumption in case it gets into the hot-water tank by mistake
Total	£1,938	

It cost a lot of money, but we had to replace our water tank, the installation was nearly 50 yards of piping away from the loft, and we bought twice the number of tubes an average family might need. I think it is more than possible to buy all that is necessary (and in that I include the controller as it makes you aware of what your system is doing) for about £700. Apart from the central-heating pump, which is proven, reliable technology, there are no moving parts, so the system should be long-lived and will pay for itself relatively quickly. When it comes to bangs for your bucks, solar thermal delivers a great return. Capital costs are affordable and the savings are significant. However, if you use professional installers it can get pricey.

We had a lot fun doing the installation;

it was like a large Ikea project. In addition to Ivan, Stephen Holroyd, a friend of the family, was staying and Jim and I roped in him and Brigit to help. It was definitely a case of many hands make light work, with people assembling the tubes and passing them up on to the roof, which saved a lot of going up and down ladders. As we finished Stephen was already making plans for his system up in Worcestershire, and deciding which part of his roof was best suited to it and easiest to get to. Even though it was mid-December, and the winter solstice and shortest day were fast approaching, we heated the water every time the sun popped through. The system is meant to be used in summer (and a bit in spring and autumn) so any winter warming is a real bonus. Roll on next summer.

Electricity from the sun and no moving parts

Solar photovoltaics, or solar PV as it is more commonly known, has been around since the 1950s, but for a long time it was only really used to provide power for satellites. Its basic building block is the photovoltaic cell, which makes direct current (DC) like a battery – hence the term 'cell' – and converts energy from the sun directly into electricity. Obviously, the power in a single one is limited, so the cells tend to be built in arrays that can provide useful amounts of energy.

Something I seem to remember from my degree over 20 years ago: PV cells are made from a crystalline structure of silicon that has had impurities added to make it a 'semi-conductor'. The energy received from the sun frees electrons that travel across the p/n junction and current is thereby generated.

You can't make your own PV array, and because of the complex manufacturing processes involved I have an innate doubt about how green solar PVs are. We have yet to invest in a large-scale array that will provide us with power for the house. Because the system produces direct current it needs to be inverted to produce the 240V of alternating current that matches the domestic supply.

SEASONAL ADJUSTMENTS

To get the maximum performance from a solar panel, thermal or PV, the sun must strike it at an angle of 90 degrees. The tilt of the Earth on its axis, and the Earth's orientation to the sun, affect the angle at which the sun's energy reaches the Earth at different times of the year. The sun is very much higher in the sky at the summer solstice than it is at the winter one. It is possible to optimize for summer or winter, or indeed for any time in between, but we find it more satisfying to adjust the angle every couple of months. This way we know we are getting as much energy as possible. It takes a little effort, but it is a reminder of the changes going on around us.

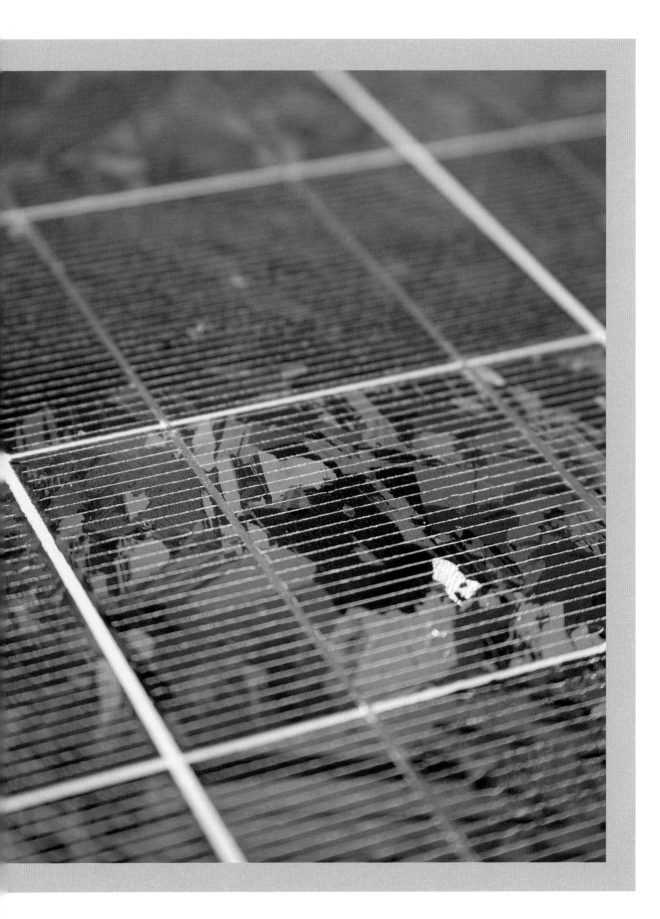

Even in Britain there is enough sunshine to make installing solar systems worthwhile

Our small arrays

We have used a couple of small arrays, which are especially convenient for locations that are a little bit remote. Our first solar panel was used to charge a 12V battery for the greenhouse heating system. It would have been prohibitively expensive to provide mains to the top paddock and we designed the system so that a small computer fan powered off the battery would provide enough air circulation in the greenhouse. As mentioned earlier, the second, slightly larger, array provides back-up charging for our pump power on the lower paddock. Both solutions are elegant, as there is no maintenance and they are completely self-contained. And they are so simple that, as

the height of the sun varies from season to season, we can easily adjust the angle at which they lean against the greenhouse or well.

Linking to the grid

It is possible to buy a complete system that allows you to be linked to the grid so that any electricity that is generated but not used can be sold back to it. Typically, electricity companies pay only about a third of what they charge per kilowatt hour, so this is not a route to an easy profit. It makes much more sense to use the energy you produce in preference to any supplied by the grid, and minimize paying for any at all!

Luxuries and Living
the Good Life

Life is all about spoiling yourself sometimes. The trick is choosing when and how

Are luxuries forbidden?

I believe that clever marketing and changes in public values have significantly reduced most people's ability to make rational decisions when it comes to spending money – and that includes me and my family. Having children probably makes it harder to resist the onslaught. Our house is full of things we don't need and should not have wasted our money on. Having said that, there are consumer goods we love to bits. I have a really clever Gaggia coffee machine that grinds the beans, makes perfect espresso, or cappuccino, and then stores the waste coffee so it can easily be added to the compost – all at the touch of a button. I don't 'need' it, but it makes great coffee and the little counter in it shows that we have made over 4,000 shots in about 15 months, so we are definitely putting it to good use. I place the coffee machine firmly in the 21st-century-living category: it is a bit frivolous, but it makes us feel that we are extremely lucky and comfortable. And there are lots more, from a very sexy Nikon digital SLR camera, to DeWalt power tools, our Mercury cooker, a cordless phone system, our (four!) Clearview wood-burners, Charlotte's recording equipment, a very large collection of CDs, DVDs and videos, my Triumph motorbike … the list is long.

It is considered normal to have more than one CD player or television in a house with teenagers and growing children. It's great to get them out of the sitting room once in a while if they want to watch something mindless or listen to some of their very long, very loud music. We have videos, DVDs, computers and all the hi-tech bits expected of us. In fact, we probably have more, as James and Charlotte have their own systems for university and I have a study full of the 'essential' IT tools. I suppose the big question is whether this is bad.

We spend a lot of time writing, researching and communicating, which means we have to, and want to, stay connected to the real world. Some people choose to live a simpler life and are able to cut themselves off. For our first four months in Cornwall we didn't have a television (the antenna came down when the scaffolding went up). I think we were so busy we didn't miss it. In fact, Brigit was in her element – she would much rather we didn't have one at all. I've never been a great one for newspapers, so the lack of television meant I was even more reliant on Radio 4 for my fix of the news. Now that we have a roof and television reception again, we can watch whenever we like, but I'm glad to say our new lifestyle has meant we usually have more interesting things to do instead.

Visiting the WEEE man at the Eden project was a sobering experience and our philosophy of replacing equipment when we wished to upgrade or if we had a problem has been superseded by trying to repair and, if an upgrade is necessary, embarking upon internal fiddling.

Electronics are a 21st-century expenditure, but we spend a lot more on other items – clothes, makeup, household goods, magazines, furniture, decoration, carpets, tools … you name it, we've probably bought it. Our move to Cornwall has at least made us much more discerning consumers and we are definitely buying fewer, better-quality goods. We also spend a lot longer looking at labels. However, having at first questioned everything, we now pretty much know what makes and types of consumables to repeat-buy.

A bit from Brigit on skincare

Just over a year ago, I picked up a copy of *The Ecologist* in a dentist's waiting room, and flicked through an article about skincare products and the potential side effects of many of their ingredients. It made pretty scary reading. The article left me feeling more than a little uncomfortable about the skincare range I had been using for the last twenty-odd years. When I got home and checked out the lists of ingredients on some of my tubs and bottles, I was shocked to discover that, apart from the occasional essential oil and herb extract, the majority of the ingredients listed had long, unpronounceable, chemical-sounding names just like the ones in the article. For example, butylphenyl methylpropional is a skin irritant and ethylhexyl methoxycinnamate is a hormone disrupter.

I knew straight away that I didn't want to continue using my existing skincare range, but I have to admit it was quite hard flushing the contents of some of the more expensive bottles down the loo, especially as I had spent most of my adult life being taken in by the 'anti-ageing', 'anti-wrinkle', 'anti-everything' claims made by the manufacturers. Now that I understand just how toxic some of the ingredients are, I'm no longer prepared to put them on my skin.

A year or so on, I now use completely natural ranges of skincare, bath and

SOME OF MY FAVOURITE PRODUCTS

Trilogy: *Rich, nourishing, especially good for dry skin, contains organic rosehip oil for renewal and repair, minimalist packaging. Tel: 01403 786053; www.trilogyproducts.com*

Weleda: *Established for 80 years, affordable, huge range including many excellent products for specific skin complaints, easy to use, easy to find. Tel: 01452 770805; www.weleda.co.uk*

Dr Hauschka: *Slightly dearer, chic packaging, holistic, high-quality products and treatments, fast delivery. Tel: 01386 792622; www.drhauschka.co.uk*

Faith in Nature: *Best soaps ever, shampoos and bath/shower products, also household cleaning products, inexpensive. Tel: 01617 642555; www.faithinnature.co.uk*

Circaroma: *Deliciously aromatic, pure and sensuous, yummy! Small company, skincare products created using aromatherapy and herbalism, slightly dearer, mail order. Tel: 020 7359 1135; www.circaroma.com*

Essential Care: *Organic range, handmade in Suffolk by aromatherapist Margaret Weed. Fantastic for sensitive, dry and allergy-prone skin. Tel: 0870 3459569; www.essential-care.co.uk*

shower products. I was worried at first that the choice would be fairly limited, but far from it. Once I had done a bit of research, testing and asking around, I realized that the range of truly natural skincare products on the market was quite extensive and that my biggest problem would be choosing which ones to use.

All of the products I now use are 100 per cent natural and contain organic ingredients wherever possible. None of them contains artificial preservatives, synthetic fragrances or petrochemicals, most come in recyclable packaging, and none is tested on animals. I did wonder whether I might age ten years overnight, but nothing like that has happened. On the contrary, I wish I had become aware and made the change years ago, because apart from the many benefits to both myself and the environment, I have developed a heightened sense of smell and am enjoying the many new aromas I am experiencing.

The lists of ingredients used in the organic and genuinely natural skincare and bathing ranges are nothing short of delicious. They include essential oils of rosehip, patchouli, chamomile, mandarin, spearmint and lavender, aloe vera, shea butter, almond and apricot oils, echinacea, green tea, brown sugar, ginger, clay earth, daisy or nasturtium extract, apple, rosemary, quince, sweet almond … A world away from the ingredients in the stuff I used to use.

So, lots of changes, and all for the better. The good news is that you don't have to spend a fortune, although some ranges are dearer than others. Nor do you have to shop exclusively in health food outlets. Some of my favourite ranges are sold in supermarkets and high-street stores. It has never been easier to try something new … and it's definitely one of the best parts about becoming greener!

Biodynamic techniques

Biodynamic techniques share the same principles and method with organic farms, but also take into account the effects of planetary influences in the growth of the plant. *Gardening for Life: The biodynamic way* by Maria Thun is a great little book if you want to find out more.

Shopping

There are goods we don't buy and shops we don't go into any more. A list would prob-ably leave me open to all sorts of libel actions, so I'll just say that multinational corporations and shop chains that are profit-driven tend not to have the well-being of the people or land that provide them with labour and raw materials fore-most in their minds. Some blatantly exploit developing-world resources and there are regular stories in the press about child labour, sweat shops and/or the rape of the planet, but things change and it is essential to ensure you are not boycotting because of old news – companies that truly clean up their act deserve to be rewarded.

Where possible we tend to try to support smaller shops and to chat to the people in the shop to see if they know where their goods come from. I'm not advocating buying everything from little hippy shops that spe-cialize in ethnic-looking frocks and smoking accessories for herbal tobacco … We have found plenty of independent 'normal' shops with lots of attractive products. Our nearest large shopping centre is Truro (St Austell, our nearest town, is going through a rede-velopment). Truro has all the shops that we now expect to find in a city centre – sadly! – but fortunately a smattering of independent shops as well. Brigit does have a tendency to home in on the more quirky shops, which may be why our needs are better catered for within the independent sector, but all our Christmas presents came from these shops and the quality and variety was great.

Hemp

Brigit found Quintessential, a small shop that specializes in Hemp products, not long after we moved to Cornwall, and I was first introduced to hemp clothing within weeks of moving to New House Farm. Hemp features lots when talking about green concerns, though usually in the form of a laugh about smoking it! The little leafy logos don't help the cause.

Luxuries and Living the Good Life

As an eco-underwear virgin,
    I was pleasantly surprised when I got
my first pair of hemp boxers

As an eco-underwear virgin, I was quite pleasantly surprised when I got my first pair of hemp boxers. Once washed and worn, they were as comfy as any cotton ones I have. Brigit has been responsible for lots of our friends, and the children's friends, being introduced to hemp. Standard presents this year have been hemp T-shirts, hemp trousers, hemp skirts, hemp socks, hemp boxers and some very nice hemp designer tops. I have no idea what Brigit is going to do next year now that hemp is no longer a novelty item in and around New House Farm.

There is a problem with many 'ethical' shops – the products tend to be more expensive than in the average high-street outlet. I suppose that's for a reason. Sharing of profits all along the supply chain or sub-optimum production methods that have not been fine-tuned to deliver maximum efficiency mean that raw materials and manufacturing costs tend to be higher.

Buzz words

ETHICAL – Supply chain from raw-material production through manufacturing to retailing does not exploit those involved. Not regulated but well-meaning; be a little sceptical.

FAIR TRADE – Fair profits paid to the producers. Regulated as a trademark, but a bit woolly as an adjective.

ORGANIC – Production adheres to the principles of the Soil Association and is well controlled.

ETHICAL SHOPPING ON THE INTERNET

There are absolutely masses of ethical, green, fair-trade sites. Anda loves Howies for clothing; it's definitely not for a crusty old fart like me. Apparently the top four here are the first clothing companies to get the official 'fair trade' mark and there is a much longer list in the resources section at the back of the book.

There is a lot of choice, and service from the internet is good, though I make sure the family always uses a credit card, not a debit or switch, as your protection against fraud is much better with a credit card. It is possible to make the case that the efficient delivery systems employed to deliver to your house can save energy when compared to delivery to a shop plus your energy expenditure in making your journey to the shop.

However, I still reckon shopping less and using public transport or walking/cycling are the best way forward. The following are our top four website shopping choices:
www.ptree.co.uk – organic and fair-trade clothing
www.gossypium.co.uk – organic and fair-trade clothing
www.bishopstontrading.co.uk – women and kids clothing
www.traidcraftshop.co.uk – many fair-trade products including clothing.

Our Christmas check list

While most people are making extensive shopping and
'to do' lists, we reduced stress by remembering there
are only three things to check off over Christmas:

- ☒ Reduce
- ☒ Reuse
- ☒ Recycle

Not bad, eh?

Christmas: the year's biggest consumer challenge

Our family Christmases have become a little excessive. We all love the festive season and tend to go a bit mad, but we agreed this time we wanted a low-impact traditional celebration. So our first Christmas at New House Farm was a challenge. The challenge was all the greater as we were very short of time. During the build-up to Christmas, filming the series was coming to its climax, the children were home with friends, and I managed to fall off a skip, split my head open and get concussion. Notwithstanding our self-imposed constraints, we did our best to have the greenest Christmas we could.

The food

Food was easy. The turkey and the sausage-meat for the stuffing were free-range and came from the local butcher's (more about the turkey below); obviously, the veggies came from the garden. We bought a Christmas pudding from a Cornish food festival at Truro, and Sophie, my gorgeous agent, sent us a lovely case of organic wine. Apart from the wine we managed to keep our food miles down by getting mostly Cornish produce. OK – it's a fair cop. I know the dried fruit for the pudding comes from far away, but Christmas without a flaming pudding is unthinkable. Perhaps next year we can try to source British sultanas and raisins …

The tree

The Christmas tree was farmed locally and we decided to go for a cut one, though it may have been better to have one with roots. We made the decision for two reasons. We could not find a nice tree with roots within a 10-mile radius and, probably more important, we have never succeeded in transplanting a tree out after Christmas – every time we have tried, the tree has died. Any advice would be gratefully received! The bonus of having a cut tree is that when the decorations come off it can immediately become a great big pile of kindling. If you want to guarantee a locally produced Christmas tree, www.christmas-tree.org.uk lists local suppliers by county and enable you to slash your tree-miles.

The presents

Shopping for presents was more problematic. We were so busy that we had to beg the forbearance of most of the family to whom we didn't send anything at all. I suppose that is green in its own right, but we do owe them, and will make it up as soon as we can. We limited the number of visits to the shops and we decided that a day each in Truro was all that was needed. It would have been more sensible to pick up presents throughout the year. We know lots of people who are that organized, including my mum, but we feel it's rather more festive to have a minor panic at the last moment.

When I had my day out, Brigit gave me a list of things for the children that she had not managed to find, and a little 'Christmas list' for herself. James and Charlotte were given instructions not to leave me alone for too long in case I got confused and wandered off. It all went well until I got home and realized the list covered both sides of the paper. My defence is that it is greener to buy less. Anyway, we had enough, and on the day everyone was extremely happy. James's present to me was particularly clever. He had found a company that supplies 'plugs' of mushroom spawn by mail order, and arranged a delivery. Now we can grow our own shiitake and oyster mushrooms on dead logs in the shade of our little no-through lane. I can't wait to see what comes up; apparently something will happen within six to nine months. On the website, www.gardeningexpress.co.uk, they sell truffle-impregnated trees that start to yield after four to six years. Is there no end to the fun?

Wrapping involved a visit to Brigit's craft cupboard. After amassing brown paper, paper carrier bags and lots of shells, beads, buttons, string and raffia, and adding a bit of creativity, the pressies all looked great. Anda made a printing block from a potato to decorate hers; they looked so professional you didn't never notice all the reused bits and pieces.

Sellotape is made from benign plant cellulose, whereas most other clear adhesive tapes are fossil-fuel- and chemical-based.

The beads and baubles (gleaned from car boot sales) that adorned the presents were duly saved after opening so they could be reused throughout the year for birthdays, and possibly make an appearance for Christmases *ad infinitum*. If you use commercial wrapping paper, it's worth knowing that you should avoid the hugely polluting glittery wrappings. Best is to use wrapping paper made from recycled paper, which is available from www.shopwwf.org.uk/store.

Open wrapped gifts carefully so you can reuse the paper, or if it is torn ensure it gets recycled. I've always ripped into presents at breakneck speed, but I have now seen the light and, apart from a couple of minor lapses, I was very grown-up this year and only a little of the paper was beyond repair.

Festive tips

We did a trawl for other ideas to reduce the impact of Christmas. You may find some applicable to you, but my best advice is to make sure you don't let guilt spoil what is a very special family time. I can unreservedly say we did not feel that being green undermined our enjoyment – we spoilt ourselves as much as we usually do. These tips are like puppies – not just for Christmas. But unlike puppies they don't wee on the carpet.

Lights
Think twice about excessive lighting displays – some people spend thousands on lights (and on the electricity bill). Brigit had always wanted white fairy lights on an outside tree. The dead cherry tree in the upper paddock provided exactly the right backdrop for our display. One tree's worth of fairy lights, powered by the waterwheel, was gorgeous. Had we paid for the electricity, however, it would have cost us nearly £20. Scale that up to some of the displays that are becoming commonplace and it all gets a bit scary. Remember that fairy lights or LED displays inside a window use a fraction of the outdoor lights' electricity. Turn them off before you go to bed or use a timer so they go off at a reasonable hour. Having them burn through the night when almost no one is passing makes little sense – except on Christmas Eve, of course.

Despite our endeavours to be green, sometimes our hearts rule our brains; Brigit has decreed that the lights look so lovely on the tree they are now to become a year-round feature – thank goodness we are getting more power from the waterwheel than we expected.

Fires
Traditional log fires using local wood are almost carbon neutral and are great to sit down by, so spoil yourself. Log fires are even better if you turn down the central heating and huddle in close. Every degree you turn it down saves you 8 per cent on your Christmas heating bill. Use the savings to buy more logs!

Oven
We believe in turkey and all the trimmings. A really simple saving is to turn off the oven about ten minutes before the turkey is due to be taken out. The oven will remain hot for at least this amount of time. Normally the oven stays hot long after you whip out the turkey, which is just wasted energy.

Going away
It doesn't make sense to me, but nearly two million Britons go abroad for Christmas, creating millions of tonnes of CO_2 in the process. Millions more will drive hundreds of miles to visit relatives and friends. If you insist on not relaxing at home, you can save

your conscience by paying to counteract the extra CO_2 you are producing. British Airways allows you to do this when you book online, or you can do so at www.climate-care.org. The easiest way to minimize the CO_2 you produce is to plant trees. Remember that it takes the entire lifecycle of six trees to counter the effects of a single long-haul flight, which means you must plant where the trees have the best chance of reaching maturity and surviving.

Wine
Find your closest organic winery by logging on to www.whyorganic.org.

Food
Christmas lunch can involve tens of thousands of food miles. Think about the wines, fruit and vegetables from all around the world that congregate in your kitchen. Over 20 million tonnes of CO_2 are emitted every year bringing the food to our UK tables. Ensure your vegetables and fruit have come from local farms by visiting your nearest farmers' market. Brigit loves visiting ours in Lostwithiel. She comes back with the most tempting array of chutneys, breads and non-mass-produced food, including Christmas cakes and puddings. Far more fun than battling the hordes in a faceless supermarket. Over $25 billion dollars' worth of pesticide is sprayed on our soil worldwide every year to help produce food. These poisons have to be extracted from drinking water, costing UK consumers millions every year. The only real way to minimize this is to grow your own or join the millions of others who are already buying organic food. Recent surveys have shown that three-quarters of UK families now buy organic food regularly or occasionally – that's really encouraging.

The turkey

I was very keen to fatten our own turkeys this year, but, with so much going on, common sense prevailed and we bought free- range from our local butcher. It wasn't cheap but it tasted magnificent. We all know that turkey factory-farming is right up there with other intensive animal-rearing practices: barbaric. Wild turkeys are naturally docile creatures. Factory-farmed turkeys are cooped up in their thousands in darkened sheds and have their beaks sliced off to prevent them damaging each other as their behaviour becomes aggressive under these conditions. If you think before you buy meat, you will probably buy less, but buy better-quality, better-reared meat. If the price is too good to be true, don't buy it.

Talking turkey

❀ The first turkey was brought to the UK by William Strickland in 1526 and sold in Bristol.

❀ Ten million turkeys are eaten in the UK at Christmas. Between 6 and 15 per cent of turkeys die during rearing in the factory sheds.

❀ Wild turkeys can fly at over 50 miles per hour.

❀ Factory-farmed turkeys are unable to fly.

❀ Turkeys in the wild sleep in trees at night.

❀ 4,200 tonnes of aluminium foil will be used to cook turkeys at Christmas. This aluminium can be either reused or recycled. Recycled aluminium uses 5 per cent of the CO_2 it takes to produce raw aluminium.

Tidying up after the party season

We thoroughly enjoy the excuse to party that Christmas and New Year provide. Having decided to throw a big New Year party to celebrate the end of our first year being green, we were very aware that this would generate sack-loads of glass bottles and aluminium beer cans.

We did our bit to minimize the waste by buying barrels of the local St Austell Brewery real ale, Tribute – what a cracking excuse! We didn't use the plastic 'glasses' offered by the brewery but used real glasses instead.

But despite our best efforts, we ran out of the correct plastic recycling sacks when we started the great clear-up on New Year's Day, and I had to make a trip to the local recycling plant as the drive was completely blocked with glass, plastic, aluminium and cardboard. We try never to make a special trip to recycle as it causes more damage than it actually saves. However, since the trailer and Land-Rover were piled high, it was probably just about worthwhile.

Get over the guilt

The more you look into what is green and what is not, the more you are aware that we live wasteful lives. There's an old Irish tease, 'If I was trying to get there I wouldn't start from here' – but the fact is we are where we are, and any movement in the right direction is worth it. Little changes

212

add up, and it's easy to make the little changes if you don't begrudge them. Enjoy the luxuries that make your life fun, but question them too.

Eco-gifts

There are an awful lot of unwanted presents around at Christmas. More important, there are a lot of unwanted presents around at all times of the year! Obviously they can be reused (via a car-boot sale or charity shop or given to some unsuspecting relative) or recycled, but it is much better to address the problem at the source. Money, vouchers and 'experiences' don't contribute to the mass of wasted resources that unwanted presents generate. A day trip out with your niece/nephew/godchildren/grandchildren will pay dividends and be remembered a lot longer than some plastic tat. Or try something completely different: an ethical gift. They are ingenious, fun and really help to make the world a better place. Your intended recipient gets a card and the knowledge that somewhere in the world some good has been done in their name.

213

ETHICAL GIFTS

Approximate cost	Present
£25	Gift of Sight for a Child – your money will be used for a simple operation to remove cataracts or to repair the damage of trachoma. Small price for a life-changing present. See www.goodgifts.org; you can also buy tools for developing-world cooperatives, seedlings, plants, bees, etc.
£80	At www.greatgifts.org, it's possible to buy a flock of sheep for a family in some of the poorest places in the world. They then have the opportunity to generate income and take charge of their own future
£30	Oxfam gives you the opportunity to fund a loo in crisis situations. Once this most basic of facilities is installed, it helps prevent the spread of disease: www.oxfam.co.uk
£25	A bicycle to help health workers in developing countries, where they could save lives by increasing healthcare delivery. We tend to be aware of, but not respond to, the fact that almost 30,000 children worldwide die each day from preventable causes. UNICEF will provide this very simple transport to help reach the remotest community areas: www.unicef.org.uk
£35	Christian Aid fishing nets provide food and an income. The organisation also do a lot more gifts, varying in cost from a few pounds to thousands: www.presentaid.org

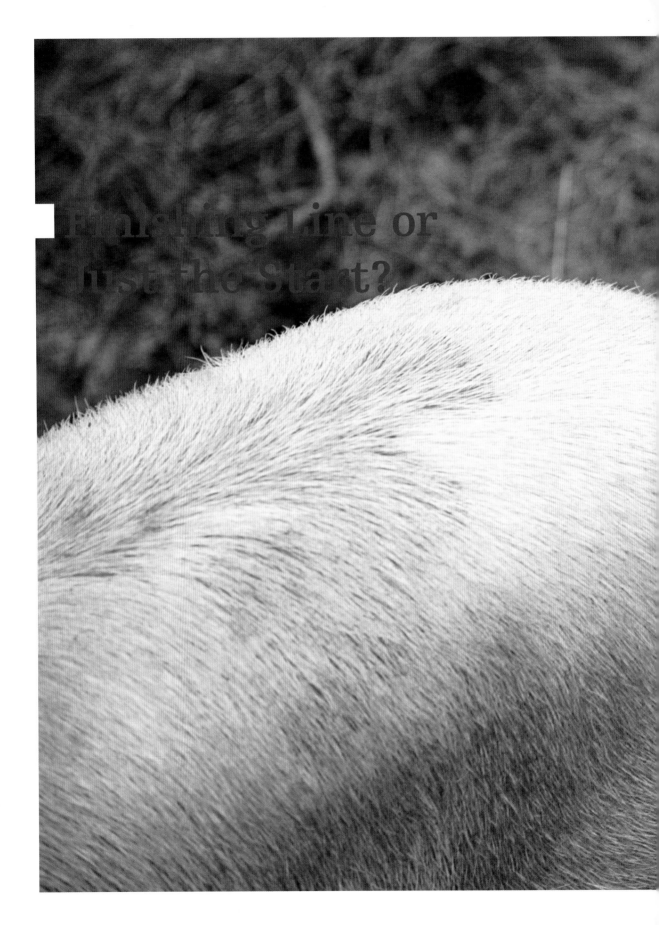

Finishing Line or
Just the Start?

Taking stock

As we look around New House Farm we can see that we
are a long way from the finishing line, but there have been
huge changes to the way we live. The house is now dry and
warm, and it contains all the mod cons needed for modern
living (water and electricity, for example!). Cosmetically, it's
a nightmare. Only a couple of rooms have been plastered
and painted, and most are just tatty. It will take a lot of
time and effort to tart them up, but there is nothing structural to sort out. In fact, the most time-consuming task will
probably be deciding what colours and materials to use
when we finish things off.

   As a family we are much more aware of green issues,
and most of our decisions are taken in the context of their
impact on the planet. Assessing where we are now shows
that we have travelled a long way, though it is hard to say

We are much more aware of green
issues and much of our decisions are taken
in the context of ther impact on the planet

whether we have met our objective: to enjoy a 21st-century lifestyle, but produce little or no waste and remove our dependence on fossil fuels. If I was to give a pithy answer to the question 'Have you achieved what you hoped to achieve?', I'd probably say: 'Nearly.' We have maintained our 21st-century lifestyle, we are producing much, much, less waste than the average family of four and our consumption of fossil fuel has been reduced to some gas for cooking and the occasional tank of diesel (if we are travelling and cannot find biodiesel).

All in all, not bad. However, we have further to go and we can see ways of reducing our impact even more. Before we left Malvern, Donnachadh McCarthy, author of *Saving the Planet without Costing the Earth*, conducted a 'green audit' on our lifestyle. The score he gave us was a respectable six out of ten – not a bad starting point. His return to Cornwall allowed us to show him what we had done to address areas where he felt we could do better. I think we followed most of his advice, but there are a few things I'm still prepared to argue over!

OUR GREEN AUDIT SCORES

Green advice	My comments	Score*
*Notes: All scores are out of 3; NYA = Not Yet Applicable, but we intend to address these points as soon as we get our act together!		
I didn't push Donnachadh for a final score, but we were a little smug even though we still we want to do a lot more.		
Convert motorbike to alternative fuel	Not happening until I find a biopetrol that won't damage my gorgeous Triumph	0
Convert car to alternative fuel	Done	3
Convert home-electricity supplier to a green electricity supplier.	Done	3
Start using bicycles for short local journeys.	Village is close so we walk, though Jim and Anda cycle	2
Use train for work journeys	We're not great at this, though we do travel by train when possible	2
Avoid car-based holidays	What's a holiday?	3
Change investments/pension to environmentally responsible financial instruments	All our funding has gone into this environmentally responsible project!	3
Ensure there is insulation underneath floors in new home before laying carpets/floor coverings	Will do if we ever get that far	NYA
Ensure opportunities for passive solar gain in new home are taken, and that external doors have insulation lobbies	Being/been addressed – we can't wait to get our 2-storey conservatory. The plans are on their way through	3
Use only organic paints and decorating materials	Not much tarting up done yet, but we have found local suppliers and will be organic	NYA
If buying new underlay for carpets, use a 100 per cent post-recycled waste product	Will do if we ever get that far – we intend to only have carpets upstairs	NYA
Avoid the use of Tetra Paks, which are currently almost impossible to recycle	Wouldn't touch them, but it is the only way to get soya milk!	1
Do not install a power shower in new home, unless it can be fully serviced by your own water and solar heating. Ensure the shower head is an aerating head	We haven't got one and we have our own water and solar-thermal heating	3 + 3
Cut the number of baths to a maximum of one per week and seek alternative means of relaxation, such as solitary meditation or yoga sessions	Number of baths reduced, even though we have our own water and heating. Relaxation?	3

Green advice	My comments	Score*
Check all adaptors/chargers, e.g. mobile-phone chargers, are only in use when needed	Checked at night before we go to bed	3
Lighting :		
* Switch off lights/heating in rooms not being used.	We switch off to save energy	
* Ensure any new light fittings can take energy-saving bulbs	Always!	
* Install energy-saving bulbs in all light fittings	Always	
* Use sensor lights for any external security lights – solar powered ideally	Done but not solar powered And we run it from a waterwheel!	3 + 3!
Install thermally lined curtains in new home – flimsy curtains allow significant heat loss	Guilty – we don't yet have many curtains. They will be thick	NYA
Ensure basic draughtproofing of all external doors/windows is carried out immediately and that all other potential draughts, e.g. through holes around piping and unused chimneys, are dealt with	Done when we replaced the doors	3
Put an air-fan above the wood-burner to ensure it heats more than just its immediate area	We are the proud owners of 'ecofans' which don't use any mains electricity — they use the fact that there is a heat differential between their top and bottom surfaces to generate their own electricity	3
Ensure nitrogen-fixing plants are included in your planting strategy for the garden	Anda and Brigit have it covered, and we plant 'green manure'	3
Take readings of all utilities on the day you move in to the new house and take that as the start date for your annual environmental audit. Monitor monthly domestic waste production on a weekly or monthly basis	Our lives at New House Farm have been all about monitoring and reducing our energy usage and waste production	3
Ensure architectural rescue items are used for new showers/baths, etc.	We reused when possible, but for high-efficiency loos/showers we went with new options	1
Plant trees to offset CO_2 emmisions caused by air travel	We've planted lots — some specifically to offset my flights to the Far East, but more than enough to compensate for them, so we are having a positive impact on CO_2 production	3
When replacing steam iron, buy a non-steam one and use a manual water sprayer.	After meeting the WEEE man we try not to replace our electrical goods. We repair instead but, to be honest, we've all but stopped ironing unless it's a special occasion	3

Planting and planning for the future

Our new lifestyle is much more in tune with nature, and we are very aware of seasonal changes and the weather. We have yet to live in Cornwall for an entire year, but all our planning involves taking account of the seasons and their effects. Even in winter, when the days start to lengthen (and are occasionally dry) we seize every opportunity to prepare for when the sap rises.

Trees
A visit that Brigit made to the Duchy Nur-sery has paid off and Tracey helped to choose the trees we planted. The family and all the friends involved in our mammoth planting session of January knew we were making a long-term investment.

The wet area in the lower paddock is now covered in willows and dogwoods. Tracey chose species with different-coloured trunks and branches, so it will look great in autumn and winter as well as when the trees are in leaf. When they establish themselves we will be able to coppice, and though we will not be self-sufficient in firewood there will be enough wood for basket work and willow weaving, and the area will provide great cover for chickens (and possibly pigs). It already looks great – and all we have is a collection of stacked and tied twigs.

The top of the garden is on its way to becoming a little wooded area. Some young trees were already growing there, and we moved ones that were too close together and planted sweet chestnut, cob, sycamore, ash, apple, damson, pear and, of course, a walnut tree I got for my birthday. This is now out of its pot and has the opportunity to flourish. I expect walnuts in four or five years, which is pretty quick, but it will be a lot longer before the tree becomes a major feature as planned.

We have opted for Cornish native species of fruit trees, rather than going for heavy croppers – a bit whimsical, but it's nice to keep old customs going. Perhaps it's years of being immersed in the history of the army, but I like tradition and I reckon years of evolution are relevant. With a bit of luck our trees will not only look good but will also provide us with a great harvest.

The growing season
Cornwall is renowned for its long growing season (mind you, we were told there was seldom a frost here!) and this year we will have the opportunity to fully exploit it. It's our chance to see exactly what can be done. Our hard work with lots of clearing and mulching, together with the rooting and clearing conducted by our first couple of pigs, has put us in a great position to start planting. It is common knowledge that you should observe your garden for a year before

you commit to reorganizing it — a principle that is completely in line with permaculture. So this year will be our first opportunity to start planning for the rest of our lives. Brigit and Anda are keen to get it right and are continually thinking through the options. We have lots of them, which is absolutely great. One thing is for sure: we are going to see many changes over the next months.

Winners and losers

Taking stock of our activities since we arrived in Cornwall allows us to learn from our experience and decide, with the benefit of hindsight, what we would do again and what we would do differently.

TAKING STOCK

Activity	Conclusion	
Hens	The hens have been a great success. We love their eggs and they are easy to look after. From a motley, bald bunch they have blossomed into healthy, plump, feathered creatures with years of laying ahead of them. They have a happy life, and even though one was killed by a predator they are a long way from the misery of a battery farm.	☺
Pigs	I love having pigs. The meat is fantastic. We have a stocked freezer and I have made bacon, sausages, hams and salamis. The pigs cleared up a couple of areas magnificently. They make use of any kitchen/food waste. They are a joy to tend to. There is a downside: Brigit and Charlotte have not eaten pork products since we killed one of them, which saddens me.	☺
Permaculture	We still have a lot to do before we can say we are following all the principles of permaculture. That said, we like the harmony we are achieving and all our long-term plans involve developing the plot further.	☺
Spring water	We couldn't afford to keep paying for our water at the rate we were using it. The spring and the wind-and solar-powered pump are a great solution. Without a shadow of doubt organizing this was money and time well spent.	☺
Rainwater harvesting	We don't do a lot, but it's really convenient. When we start to develop the outbuildings (which are further away from the spring than the house) it will become even more important. Good savings for a little outlay.	☺
Whole-house ventilation	We had major damp problems and lots of conflicting advice. Our solution uses wood-burners and a system that dries the air and transfers warm air around the house, and is a winner. The property is drying out very effectively, and the air has been cleaned significantly if the state of the filters is anything to go by. We have not achieved sufficient heating upstairs to make it 'warm', but we were asking an awful lot of a	☺

Activity	Conclusion	
	heat-recovery system. Once we have supplemented the system with radiators running off the back boilers on the wood-burners it may make all the difference.	
Wood-burners	We love them. They are the best we could find. They do a cracking job. They use quite a lot of wood. We have yet to source free wood that we can store. That's not because there is a shortage in Cornwall; it's just that we haven't got our stockpiling act together.	☺
Insulation	Having experimented with recycled textiles and sheeps' wool, we are starting to retain some of the heat we put into the house. I think it is fair to say we prefer the wool from upland sheep as we are using something that would otherwise be thrown away. That said, both products are competitive in the insulation market so why go for a conventional, non-green solution?	☺
Greenhouse heating	Lots of frost outside, none inside. Need I say more! The location for the greenhouse is excellent and we have the advantage of a south-facing wall, but our little solar panel, a computer fan and a lot of 'imploded' glass is a winning combination.	☺
Solar thermal	Bangs for bucks, this seems to be a solution everyone should invest in. We haven't yet seen how it performs in spring/summer, when it is due to come into its own, but even in winter it responds to a sunny day which can't be bad!	☺
Compost	Brigit swears by it, Anda loves doing it and it makes an awful lot of sense to me. We need compost and we make it by the heapful. It is the only way to process organic waste that you don't give to animals.	☺
Salads	We had a great summer. We were eating our salads within about five weeks of getting here. We still have winter salads in the greenhouse which we have been picking and eating through January and February. It's a no-brainer; everyone should grow salad.	☺

I can't believe that we spent so many years paying £1.99 for a little bag (filled with some form of inert/sterilizing gas) when Brigit can go out to pick a colander of salad and regularly come back with 15–20 varieties of leaves and flowers. I'd never heard of lots of them (e.g., 'chop suey greens' – guess what – they actually taste of chop-suey!) and I bet I'm not alone. We fed the 5000 on our salads during the summer months, and have even got to the point where we seldom find creepy-crawlies in them after they have been washed.

I have to put in a special mention for our cucumbers. I've never grown them before, and thought they needed to be under glass or mollycoddled. Not so. We had dozens, and they were tasty and huge.

To be honest, the tomatoes were a real disappointment. I know we were late in planting, and they had no cover, but from the masses of really exciting types we had very few ripe ones, and only a couple of dozen jars of green tomato chutney. Next year …

To end on a high note, our radishes were huge and tasty (actually, they were seriously hot). Think of a big red swede, then think of something slightly bigger, and you are getting close to the size of our radishes. And they were organic and definitely not modified!

| Squashes | We had brilliant courgettes, patti-pans and all forms of squash. There was a bit of a glut ☺ during late summer, but the great thing about squashes and marrows is the way they keep. We were pleased that the different types tasted so different – acorn was the overall favourite. Our staple diet during winter is warm soup at lunchtime. Roast squash with chilli and garlic has got to be one of the best. |

| Vegetables | Gardening in an organic way has its problems. The caterpillars had their share of our ☺ cabbages (red, white and green) and brussels sprouts and everything else leafy. Luckily, there was still enough for us — just. We also had some very tasty leeks, broccoli, turnips, parsnips, artichokes, beans (green, runner and French), beetroot, celery, spinach, kale … Our vegetable patch looked truly vibrant and I loved the variety of plants. Next year we will definitely keep the diversity, but we intend to grow more of each individual type so that we can sell the excess. |

| Fruits | We had a couple of apples off the trees we inherited, and found masses of wild ☺ strawberries which were so small they were only good for nibbling as you walked around the garden (not to be scoffed at). All the soft fruits (strawberries, raspberries and blueberries) that we bought and planted will hopefully come into their own next year. |

I have a cunning plan for a composter that will act faster and hotter than anything we have to date. It involves a chest freezer and a car axle!

Transport – biodiesel	Not having petrol-driven cars was not that difficult, and we now have a very economical, green solution to our transport needs. Making our fuel takes time, and we are always on the lookout for used vegetable oil, but it works. Hopefully it won't be too long before biodiesel is widely available. ☺
Reducing	Changing our shopping pattern has not been as hard as we might have thought. Bulk delivery of organic food, and food from the garden, has helped. We still enjoy a trip to the shops (it's an excuse for a coffee or lunch out). However, we automatically check the origin of what we are buying and tend not to requent the large chains whenever possible. ☺
Reusing	We keep things that we have a use for, so we are reusing wherever possible. Brigit has boxes (I'm talking packing, not shoe, boxes) full of beads, buttons, materials and 'crafty' things, all being saved for when we have time to reuse them. Anything we have no further use for tends to go to a charity shop, car boot sale or for recycling. ☺
Recycling	We have fully embraced recycling. It does take a little more time but, once organized, it can be done! ☺

Lots of smiley faces, which is not bad for our first year. We have learnt lots and have every intention of doing much more of everything this year.

What happens next?

New House Farm is now a thriving home and smallholding. Diversity is the order of the day, partly to ensure that any failure is only a little failure, but partly to fully exploit the limited amount of land available to us – you never know, one of our new ventures may just be the most lucrative, or most successful, one. All of us have ideas about what we'd like to do, so here, in no particular order, are some of the things we want to have a go at:

* Historically, there have been orchards at New House Farm for hundreds of years, and we need even more fruit trees, including exotics like peaches against the walls in the top paddock. I still haven't planted out the two fig trees I got for my 40th birthday. We have to choose exactly the right spot.
* I'd like to make cider in the 'cider shed'.
* I want to build a smoker again.
* Lots more soft fruits are needed for preserves.
* I have to plant mushrooms in the logs in the lane.
* I have a cunning plan for a composter that will act faster and hotter than anything we have to date. It involves a chest freezer and a car axle!
* We could do with harnessing the power from the lower paddock. I quite like the idea of a water turbine as we haven't used one yet.

* I've seen a 'green' rickshaw that would be great for local trips. We need to buy or make one.
* Brigit has seen a plan for a reed-bed swimming pool.
* We have yet to address the house's cosmetic issue. We need paints, wallpaper, furnishings, carpets and all the trimmings.
* We have to decide how to take a green holiday. I fancy building an eco-gypsy caravan.
* What about keeping a steer for beef? We have access to the necessary land, and I fancy an old breed like an English Longhorn, slow to mature but very special to eat.
* Keeping a couple of lambs may be useful.
* We'll definitely fatten up turkeys, and possibly some geese, for Christmas.
* Instead of buying chicken meat I am keen to keep 'chunky chickens'. The ones kept for supermarkets tend to be rushed

226

and may be slaughtered when they are as young as eight to ten weeks. We will probably wait a lot longer, and the chickens will definitely range freely. I quite fancy having capons, but have no idea how to sex the chicks and whip off the necessary bits.

* We are already fattening more pigs. The cycle will continue.

* I will definitely get a hive and start keeping bees again. Mead will be the drink of choice next Christmas!

* Ever since I saw Tom and Barbara building a 'biogas' generator in their cellar, in *The Good Life*, I have wanted to make my own biogas. I will have to source enough pooh, but its definitely worth a go.

* When fuel was in short supply during and after the Second World War, 'gasifiers' that extracted combustible gas from organic material like wood were developed to run conventional four-stroke combustion engines. I reckon it will be possible to run a generator, or even a vehicle, with one. Like the biogas system, it will have to be a long way from the house.

* A biodiesel generator would provide any power we need that doesn't come from sustainable sources. Excess electricity could be sold to the grid.

* A grid-linked solar PV system would reduce our reliance on the grid and we would sell back what we don't use.

* We need to earn enough to live, so: we will be attending farmers' markets to

sell excess produce. We also intend to build an 'honesty cart' for the top of the lane so that people can take what they want and pop the cash into a sealed box.

* Brigit and Anda reckon flowers are a real money-maker.

* We intend to run 'green' courses in the outbuildings. We will have to decide whether we should renovate the necessary barns or whether Brigit will finally get a yurt/earth shelter that we can use as a classroom. There is lots we want to teach and share, but to begin with we will have to decide what to concentrate on. The whole endeavour will have to be green, so it will need to be self-powering/heating (www.itsnoteasybeinggreen.org).

* Although it is early days in our coppice, Brigit will be able to use the skills she has learnt and make willow baskets.

* Our spring water is fantastic, and once we sort out the abstraction licence etc. we may bottle it for sale locally as a very exclusive product, with no transport miles.

* We have a cunning plan to extend our interests to the local area and invest in a sustainable source of energy that will power a lot more than just New House Farm. Initial investigations are promising – enough said, as the scheme has to remain confidential.

* There is more wind power to be captured. We have to address planning for Turbine 3.

* To increase our work space we will renovate some of the outbuildings. We intend to

put sedum on at least one of the roofs. (It's a small, truculent plant that grows thickly but is not very high, and it makes a great roofing material – a bit like very posh turf!).

* We need a polytunnel so that we can nurture the more delicate plants we want to grow. Jim and I found plans on the internet for one made from plumbing waste pipes. It sounds so ridiculous that we have decided to build one.
* We have yet to use our 'grey water' (from the bath, sink, shower, washing machine).
* We are going to dig out a pond. Old plans and maps show there was one to the north of the house for hundreds of years. It will be great for wildlife, and I fancy trying to introduce British crayfish (not the nasty North American signal crayfish). Tasty.

228

This is a huge list, but it is by no means complete. When one of us sees something they fall in love with it goes on the list, and its priority depends on how much it costs and how much we want it. One of the beauties of having the space and the inclination is that anything is possible.

Conclusion

I am the luckiest man I know. I have a gorgeous family. I live in a magnificent old farmhouse in a lovely village near the seaside. I eat and drink quality food, wine and water. I have all the trappings of modern living available to me. I do what I love for a living. I'm never sure what tomorrow will bring. Not everyone can have all that I have – they are probably not that lucky – but everyone can have some of it.

I'm the luckiest man I know.
Not everyone can have all that I have,
but everyone can have some of it

Acknowledgements/Gallery

Somehow we failed to get pictures of all our helpers, partly due to some being camera shy and partly through incompetence on our part.

As you can see, we had the help and support of some very fine people.

Thank you.

Resources

About this book

The paper used for the inside pages of this book has been independently certified as coming from well-managed forests and other controlled sources according to the rules of the Forest Stewardship Council.

This book has been printed and bound by Butler & Tanner Ltd, an ISO 14001 accredited company.

Further reading

One of the great perks about launching into this project has been the growth of our library. We love books but on many occasions we have been unable to justify a purchase because it is too similar to another book we already own or the information is available on the Internet. However, we decided that our longer-term aim would be to run courses and share our findings, so we now have a library for people to browse through. When we get our act together, you will be able to check out our websites for information and links to lots of other green sites:
www.itsnoteasybeinggreen.org
www.greenwebpage.com

Books

A Good Life
Leo Hickman (Eden Books)
It makes you think.

A Greener Life
Clarissa Dickson Wright and Johnny Scott (Kyle Cathie)
A very easy read for pro-field-sports meat eaters, not for vegans.

Bring Me My Bow
John Seymour (Turnstone P)
A classic.

Change the World for a Fiver
(Short Books)
A worthwhile investment.

Compost
Clare Foster (Cassell Illustrated)
You can never get too much information
about compost.

Do the Right Things
Pushpinder Khaneka
(New Internationalist Publications)
One of Brigit's must-haves.

Dowsing
W.H. Trinder (British Society of Dowsers)
We used it.

Encyclopaedia of Organic Gardening
(Organic Organisation)
Lots of information.

From the Fryer to the Fuel Tank
Joshua Tickell (Eco-Logic Books)
A useful guide to making and using biodiesel. It also covers the use of 'straight' vegetable oil as a fuel.

Go Make a Difference
(Think Publishing)
Ways to save the planet; a lot of useful
contacts.

Home Curing of Bacon and Hams
(Ministry of Agriculture)
A Second World War pamphlet that we
were given by James Cooper – a great read.

Home Sausage Making
Peary and Reavis (Storey Books)
Great and lots of fun.

Home Smoking and Curing
Keith Erlandson (Ebury Press)
Old, but I continue to dip into it.

How to do Just About Anything
(Readers Digest)
The title says it all! A bit dated now.

It's a Breeze
Hugh Piggott (Centre for Alternative
Technology)
A guide for complete beginners on how to
choose a small, commercial wind turbine.
We found it quite expensive for what it is.

Lifting the Lid
Peter Harper and Louis Halestarp
(Centre for Alternative Technology)
Everything that you will ever need to know
about composting loos.

Not on the Label
Felicity Lawrence (Penguin)
Will drive you away from processed food.

Saving the Planet Without Costing the Earth
Donnachadh McCarthy (Fusion Press)
Do the test and see what you should be
doing.

*The Complete Book of Raising Livestock
and Poultry*
Katie Thear and Dr Alistair Fraser (Pan)
Good general information and helps in
making decisions about what to have.

The Collins Complete DIY Manual
Albert Jackson and David Day
(HarperCollins Illustrated)
Jim and I dip into this all the time to check
what we should be doing!

233

The Earth Care Manual
Patrick Whitefield (Permanent Publications)
A fantastic book on permaculture.

The Ecology of Building Materials
Bjorn Berge (Architectural Press)
We dipped, but didn't use.

The Energy Beyond Oil
Paul Mobbs (Matador)
Interesting.

The Organic Salad Garden
Joy Larkcom (Frances Lincoln)
One of Brigit's favourites

The (New) Complete Book of Self-sufficiency
John Seymour (Dorling Kindersley)
234   I've had the original version for years.
It's great.

The Rough Guide to Ethical Shopping
Duncan Clarke (Rough Guides)
Another of Brigit's specials.

The Solar House
Daniel D. Chiras (Chelsea Green Publishing)
Good background on solar heating and
cooling.

The Ultimate Natural Beauty Book
Josephine Fairley (Kyle Cathie)
Definitely worth having.

Vegetable Gardening
(Murdoch Books)
It does what it says on the title.

Magazines

There are a number of good monthly and
quarterly magazines with up-to-the-minute
news, information and advice on all things
green.

Ethical Consumer
Promotes change by informing and
empowering the customer.
www.ethicalconsumer.org
Tel: 0161 226 2929

Living Earth
The Soil Association magazine.
www.soilassociation.org
Tel: 0117 314 5000

Permaculture Magazine
Solutions for sustainable living; colourful,
informative and inspiring. It's a great
magazine and its were Brigit found
details of the course she did with Patrick
Whitefield. Definitely worth looking at!
www.permaculture.co.uk
Tel: 0845 458 4150

The Ecologist
It's been around for a long time and is
full of information.
www.theecologist.org
Tel: +44 (0)20 7351 3578

We also recommend using the Internet,
obviously, and word of mouth.

Useful contacts

Chapter One: We're Off!

Andy Hopper
www.andrewhopper@andrewcharles.net;
tel 020 7976 4348

Aluminium Roofline Products (ARP)
www.arp-ltd.com; tel: 01162 894422

Archangel Yurts
tel: 01749 890457

Ardesia Max Slates – EBC UK Ltd
www.e-b-d-uk.com; tel: 01909 479276

Colin Marshall
tel: 01726 61221

Ecotricity
www.ecotricity.co.uk; tel: 0800 0326 100

Ledbury Removals
tel: 01531 633322

Forrester Roofing
tel: 01726 64142

Gilbert and Goode Builders
tel: 01726 64800

Woodland Yurts Ltd
www.woodlandyurts.co.uk; 01275 879705

Chapter Two: Reducing Waste

A general good website for information
about general recycling, symbols and all
sorts of environmental links is:
www.wasteconnect.co.uk

Chapter Three: Self-sufficiency

David Bailey (Walter Bailey Par Ltd)
tel: 01726 812245

Greenhouses UK
www.greenhouses-uk.com; tel 08456 449394

Jane Howarth – Battery Hen Welfare Trust
www.thehenhouse.co.uk; tel 07773 596927

Krysteline
www.krysteline.net; tel: 08706 000033

Chapter Four: Water Water Everywhere

FRANK
www.frankwater.com

Hydra SW Engineering
tel: 01726 862000

PSL (Southwest) Ltd – metal products
tel: 01208 75677

Travis Perkins (nationwide – our local
is in St Blazey)
tel: 01726 817171

Par Batteries and Brakes
tel: 01726 812686

Wateraid
www.wateraid.org; tel: 020 7793 4500

Chapter Five: Fuel and Travelling

Sources of biodiesel in the UK:
www.biodieselfillingstations.co.uk

Caframo Ecofan
www.caframo.com

Clearview Stoves
www.clearviewstoves.com;
tel: 01584 878100

GMO Cars Ltd
tel: 01736 331011

Ubbink houses
www.ubbink.co.uk; tel: 01280 700211

Second Nature
www.secondnatureuk.com;
tel: 01768 486285

Sustainable Technology Ltd
www.ukhornets.co.uk

Chapter Six: Blowing Hot and Cold

British Wind Energy Association
www.bwea.com

Navitron
www.navitron.co.uk; tel: 08707 401330

Office of the Deputy Prime Minister
www.odpm.gov.uk

Proven
www.provenenergy.com

Chapter Seven: Luxuries and Living the Good Life

Circaroma
www.circaroma.com; tel: 020 7359 1135

Dr.Hauschka
www.drhauschka.co.uk; tel: 01386 792622

Essential Care
www.essentialcare.co.uk; tel: 08703 459569

Faith in Nature
www.faithinnature.co.uk; tel: 01617 642555

Fragrant Earth
www.fragrant-earth.com; tel 01458 831 216

G. Baldwin & Co (Aromatherapy oils, herbs and waxes)
www.baldwins.co.uk; tel: 020 7703 5550

Laughing Bird
www.laughingbird-bodycare.co.uk;
tel 01248 600 034

Organic Blue
www.organicblue.com; tel: 020 8206 2066

Spiezia
www.spiezia.co.uk; tel: 08708 508851

The Devon Meat Company
www.thedevonmeatcompany.com;
tel 01548 821 034

Trilogy
www.trilogyproducts.com;
tel: 01403 786053

Weleda
www.weleda.co.uk; tel: 01452 770805

www.climatecare.org
www.gardeningexpress.co.uk
www.goodgifts.org
www.greatgifts.org
www.oxfam.co.uk
www.presentaid.org
www.shopwwf.org.uk/store
www.unicef.org.uk

Wind turbines: neighbours and planning

Government policy on wind turbines is set out in Planning Policy Guidance Notes 22 'Renewable Energy' (PPG22). This is soon to be replaced by a new Planning Policy Statement 22, which is currently available as a consultation draft. PPG 22 was published in 1993 and is therefore somewhat out of date. Its approach is generally to support renewable energy, but it identifies the following factors as needing to be investigated before planning permission is granted:

Planning permission check list:

• Proximity to power lines, airports roads and railways
• Shadow flicker
• Noise
• Electromagnetic interference
• Siting and the landscape
• Ecology, archaeology and listed buildings
• Disturbance during construction

The document's main emphasis is on large-scale commercial turbines, but small-scale ones are subject to similar considerations. It is best to address all the points that can be raised in protest when you apply for permission, as this can stop a lot of hares running. The new consultation draft, PPS 22, is more up to date and more supportive of all forms of renewable energy. It sets out visual impact, noise and national designations (such as areas of outstanding natural beauty, and wildlife and archaeological sites) as issues to be considered. Copies of both documents can be viewed on the Office of the Deputy Prime Minister's website.

In every local government area a Local Plan provides an interpretation of national policy, and planning applications are determined in accordance with it unless material considerations indicate otherwise – the issues in PPGs and PPSs can be material considerations.

237

Index

Page numbers in *italics* denote photographs or diagrams.

238

240